Temporal Topos Methods for the Philosophy of Natural Sciences

Goro C. Kato

Temporal Topos Methods for the Philosophy of Natural Sciences

The Nexus of Ontology and Epistemology

Springer

Goro C. Kato
Professor Emeritus
Cal Poly State University
San Luis Obispo, CA, USA

ISBN 978-981-96-2419-5 ISBN 978-981-96-2420-1 (eBook)
https://doi.org/10.1007/978-981-96-2420-1

This Springer imprint is published by the registered company Springer Nature Singapore Pte Ltd.
The registered company address is: 152 Beach Road, #21-01/04 Gateway East, Singapore 189721, Singapore

Thus, the task is not so much to see what no one has yet seen, but to think what nobody yet has thought about that which everyone sees.

Sometimes translated as:

So, the problem is not so much to see what nobody has yet seen, as to think what nobody has yet thought concerning that which everybody sees.

Arthur Schopenhauer in his two-volume work Parerga und Paralipomena

I dedicate this book to my father, the late Takeo Kato.

Preface

Before we begin our story with the rigorous language of mathematics, especially for the readers with a humanities background, we shall take a peek at the stage behind the curtain: what is a category? A category corresponds to a stage where a drama is going to be performed. Then, what are the characters on stage? The characters correspond to objects of a category, i.e., objects are the theater-characters on stage.

So far, we have characters (objects) and a stage (a category) for the drama. This drama is a story about temporal topos theory. The next pertinent question is, "How do characters interact?" Their interactions correspond to morphisms among characters. Then as the story develops, how do we express the changing relationships among characters? Namely, we need a device to express the flow of the drama on stage as time progresses. The evolution of the story is handled by the mathematical notion of temporal site. Sometimes we need to focus on an individual character's speech, but sometimes we need to understand the influences of all the characters on stage. This process is a bridge from an individual to the whole. The bridge corresponds to the notion of a sheaf connecting the individual to the whole. The title of the play is Reality.

Shall we raise the house curtain now?

A treatise on philosophical topics written by a mathematically trained author has potential to please philosophers and mathematicians through an optimistic view. On the other hand, pessimistically speaking, such treatises please neither exact scientists nor humanity specialists. The approach taken is that we explicitly indicate what is not meant for exact scientists and what is meant for humanity specialists.

We use concepts developed in categories and sheaves as a language to describe our theory called temporal topos. Whenever one tries to apply such concepts from pure mathematics fields, we have to modify established theories to express our thoughts. A well-known case of modification of a theory is Einstein's general theory of relativity where Riemannian geometry had to be modified to better fit our physical universe.

Our goal is to provide universal formulations and views in terms of mathematical concepts in categories and sheaves for the following matters:

(A) to formulate the notion of a process emerging as a global entity from local data,
(B) to delimit and accentuate the difference between an animate and an inanimate entity,
(C) to cast the fundamental notions of quantum physics and relativity, and
(D) to provide unified views of and perspectives on Western and Eastern philosophies.

Note that topics (A), (B), (C), and (D) correspond to Chaps. 2, 3, 4, and 5, respectively.

San Luis Obispo, CA, USA
August 2024

Goro C. Kato

Acknowledgments

My most sincere gratitude goes to collaborator Prof. Hajime Sugio at Sophia University in Tokyo. Dr. Sugio has written parts of Chaps. 3 and 5 outlining the background. The wealth and depth of his knowledge in philosophy of sciences have provided productive and pleasurable hours for the author. I thank Prof. Kazuo Nishimura for valuable conversations during our earlier works. It is my pleasure to acknowledge the communication with Prof. Menas Kafatos, Prof. Sisir Roy, Dr. Neil Theise, especially in Eastern philosophy, and with Prof. Daniele Struppa in life philosophy. Many discussions with Prof. Fred van Oystaeyen at the University of Antwerp have been fruitful by comparing his theory with the theory of temporal topos. The communication with poet and physicist Dr. Ioannis Raptis has been academically significant, and I am sincerely grateful to Dr. Raptis for reading through my final manuscript and providing very constructive comments. Professor Karl Erich Wolff encouraged me to extend the notion of a temporal site to a more generalized form at the early stage of temporal topos development. The author has enjoyed long conversations with Dr. Wolff on German philosophy during his visits to Darmstadt.

Even before this project began, Springer Senior Editor Ms. Juno Kawakami and I began communicating about philosophy. I am also grateful for her encouragement for this treatise and her patience allowing me to extend this project's deadline three times. I would also like to thank Mr. Hiroshi Nozaki for his superb figure editing.

Christine Lane Willis, my wife, has kindly proofread the entire manuscript correcting my English usage, for which I am deeply grateful.

I also acknowledge the following universities and research institutes for their invitations: Kyoto University, the Research Institute for Mathematical Sciences of Kyoto University, Osaka University, Waseda University, Shizuoka University, Okayama University, Tsukuba University, Josai University, Sophia University, Tohoku University, Kanazawa University, the Institute for Advanced Study (Princeton, NJ), Antwerp University (Antwerp, Belgium), Imperial College (London, UK), Lund University (Lund, Sweden), Darmstadt University (Darmstadt, Germany), Calabria University (Calabria, Italy), George Mason University (Fairfax, VA), Indian Statistical

Institute (Kolkata, India) and the Rumanian Academy of Sciences (Bucharest, Rumania). Finally, I thank the Mathematics Department and the College of Mathematics and Sciences of California Polytechnic State University in San Luis Obispo, California, for its support.

Contents

About the Author

Goro C. Kato is a professor emeritus of mathematics at California Polytechnic State University, San Luis Obispo, CA, USA. Kato was born in 1948 in Kariya, Japan, and he came to the United States on a Rotary International Foundation Fellowship. He earned his Ph.D. in 1979 from the University of Rochester, Rochester, NY, and his Ph.D. thesis advisor was Professor Saul Lubkin. His areas of study are p-adic cohomology (ref: *The Heart of Cohomology* published by Springer-Verlag, 2006), the theory of D-modules (ref: *Fundamentals of Algebraic Microlocal Analysis*, with coauthor Daniele Struppa, published by Chapman & Hall/CRC, 1999), and temporal topos theory (ref: *Elements of Temporal Topos*, published by Abramis Academic Publishing, 2013). Kato's memoir on his visits from 1986–2008 to the Institute for Advanced Study, Princeton, NJ, was published (in Japanese) by Iwanami-Shoten, Tokyo, 2019. He has given talks in Japan, Belgium, Italy, the UK, Sweden, Germany, the Netherlands, Romania, India, and the USA.

Chapter 1
Introduction

Contents

... and he (Kant) tries, by distinguishing between the phenomenal (%) *and the intelligible world* (#)*, to give a transcendental solution of the conflict between natural causality* (%) *and freedom of will* (#)*. ... Any reasonable interpretation of Kant's attempt in the framework of Laplacian physics seems to require the existence of the individual from eternity to eternity, either in the form of a Leibnizian monad, or with the Indians and Schopenhauer, by virtue of metaphysics. ....*

Herman Weyl, in *Philosophy of Mathematics and Natural Science*, Princeton University Press, 1949, page 210.

(We will give the comments for (%) and (#) later.)

In our everyday life, as we look around when taking a walk, we see the scenery changing; closer trees and faraway mountains are moving as we expect according to our experience with distance and time. This is a macrocosm phenomenon. We can intuitively capture the macroscopic world reasonably well since we were born into the world. However, if you want to observe an electron in microcosm, i.e., in a microscopic world, it is a completely different story as Grothendieck comments on the difference:

Changing from classical physics to quantum physics is like from French to Chinese. .... Strangely, against everything, 'clouds of probability' remind me of

G. C. Kato, *Temporal Topos Methods for the Philosophy of Natural Sciences*,
https://doi.org/10.1007/978-981-96-2420-1_1

hard-to-grasp open neighborhoods, filled in topos, to surround an imaginary 'point' with strongly attached imagination ....

in his essay *Reaping and Sowing*, Reflection and testimony on a past as a mathematician, English Translation by Childe, version July 18th, 2022.

1.1 Prolegomenous Remarks

Sunshine with a spring breeze, enters temporal topos with an initial object and a terminal object

Reality Act I, Scene i.

When we try to see the microcosm directly, sometimes also called a subatomic world, first we should be aware that *we* are not part of the subatomic world; on the other hand, such a subatomic world is a part of the macroworld. Namely, the microcosm is contained in the macrocosm. Secondly, not only 'we' as an observer, but also a device to observe or measure such a small object like an electron is also an object belonging to a macrocosm. Namely, we should say:

When we are not in Rome, don't expect us to behave as the Romans do.

We human beings do not behave as subatomic creatures do. However, we have an impressive physics tool as an extremely accurate device called quantum mechanics to deal with the subatomic world.

An electron inside a rock and an electron in a protein molecule inside a brain are the same. Even though for some fractals, where *the local equals the global*, knowing the local is equivalent to knowing the global; however, the emerged global objects from local subatomic entities are often quite different from an individual local object. We need to know the process or mechanism for those simple subatomic entities being organized to reach a global entity, e.g., a brain. This quest is one of the main motivations for writing this monograph.

That is, our treatise is written for readers interested in unifying philosophical views of sciences with the help of mathematics to make assertions precise and exact.

One of our goals is to give a unifying category-sheaf theoretic formulation for the process of organizing locally trivial entities like atoms to the emerged global entity like a human, which is capable of self-awareness.

Our model is simple: three concepts of S, O, and q, expressed as

$$q : \mathrm{S} \to \mathrm{O}, \qquad (@)$$

where S is a *temporal site*, O is a codomain or target category having an initial object and a terminal object, and q is an ur-functor. All these concepts will be explained in what will follow.

Note that an object of S plays the role of a parameter called a *generalized time period*, even for the usual time concept itself.

We will devote the following chapters to explaining the interplay of these concepts. The totality S^ of all those q's as in (@) is said to be a *temporal topos* over a temporal site S.

Another goal of this treatise is to provide the interconnections among physical sciences, cognitive sciences, and philosophies expressed in terms of concepts based on temporal topos S^. Such unifying mathematical ideas are provided by category theory and sheaf theory which belong to modern mathematics mostly developed after WWII. Note that the first paper on categories was published by Samuel Eilenberg and Saunders MacLane, and sheaf theory was originally introduced by Jean Leray and Kiyoshi Oka into topology and several complex variables, respectively, in 1950's.

In mathematics, we begin with definitions and axioms as foundations. All the assertions must be consistent with axioms, i.e., having no contradictions within the proposed theory.

In mathematics, it is very important to distinguish the notion of an *equality* '=' from the notion of an *isomorphism* '$\approx$'. That is, we distinguish the difference between the equality, i.e., *being the same* and *being isomorphic*, i.e., *being essentially the same*. We must and will make it clear whether there exists a unique object or there exist isomorphic objects. In the above, even though we mentioned that an electron in a rock is the same as an electron in a protein molecule, those two electrons are in different places. Hence, they are only structurally the same elementary particles. Namely, they are not equal, i.e., ' $\neq$ ', but are essentially the same.

How do we express the notion of essentially the same-ness?

Our answer to this question is expressed in terms of the categorical notion as follows.

Use the notion of an isomorphism of functors.

We need to explain what a functor is and what an isomorphism is to give a precise meaning to the above answer, which will be done in what will follow and in the Appendices.

To be more precise, let us use an expression from mathematics even though the definitions of a functor and an ur-functor are given later in the Introduction. Let us consider n electrons which we denote as $\{E_j\}_j$ or simply as $\{E_j\}$, where j is an element of an index set $J = \{1, 2, \dots n\}$, i.e., $j = 1, 2, \dots, n$. Then let $\{e_j\}$ be the corresponding (ur-)functors as the 'avatars' of $\{E_j\}$. When isomorphic objects $\{e_j\}_j$ are given, one can consider the class $[e_j]$ of all these isomorphic objects $\{e_j\}_j$. Then a class $[e_j]$ (which is called an equivalence class) becomes only one object. Then the class $[e_j]$ is the unique electron ur-functor representing all the isomorphic electrons $\{e_j\}_j$. Note that isomorphic objects e_j and e_i, where $j \neq i$, can have different states, i.e., e_j can be in an ur-particle state e_j (V) for an object V in temporal site S, and e_i can be in an ur-wave state. See what follows for the meaning of these conclusions.

Note Such a functor e_j is a *contravariant* ur-functor, i.e., a ur-presheaf, and furthermore, e_j is an ur-presheaf. See Appendix I for the concept of an equivalence class.

Categorically speaking, two special objects α and ω in the following diagrams are an initial object and a terminal object of a category $S^{\wedge}$, respectively.

The definitions, characterizations, and properties of those special objects are as follows. (See Appendices for the definitions.) For any object q in $S^{\wedge}$, there exists an arrow, called a morphism, from an initial object α to q. Furthermore, such an arrow is unique, i.e., only one such arrow exists from α to q. For a terminal object ω, things are dualistic, namely, for an arbitrary object p in category $S^{\wedge}$, there always exists a uniquely determined arrow (a morphism), from p to ω.

In what will follow in this treatise, a terminal object ω will correspond to spacetime ur-presheaf, and an initial object α corresponds to the One, the Ultimate, Brahman, …, which depends upon various schools of thoughts, philosophies, religions in the East and the West.

History: Pure mathematics has a long history, and the elementary notion of arithmetic has an even longer history. The crucial concepts in mathematics were discovered and developed in the nineteenth century by mathematicians like Gauss, Cauchy, Riemann, Galois, etc., and the sophisticated developments in the twentieth century were made by, e.g., D. Hilbert, H. Weyl, C. L. Siegel, H. Poincare, A. Weil, J-P. Serre, A. Grothendieck, P. Deligne, M. Sato, etc. After World War II, a shift toward more abstract mathematics was first initiated by a mathematician group called *Bourbaki* in France. Further abstract modern mathematics was accelerated in the study of algebraic geometry by Alexandre Grothendieck.

Our mathematical language used in this treatise is coming from categories and sheaves. Sheaf theory and sheaf cohomology theory connect the local data and the global object (which began with H. Cartan's seminars in the 1950's for the study of the topological aspects of holomorphic functions in several complex variables).

1.2 Big Picture

Let us begin with the following symbolic diagrams (ω) and (α)for our program. Our goal is to construct a theory represented by horizontal and vertical arrows in the following two diagrams (ω) and (α). Namely, they are the directions from the small to the large. This is the direction of descent, i.e., obtaining a global object by gluing local trivial data.

For a terminal object ω of $S^{\wedge}$:

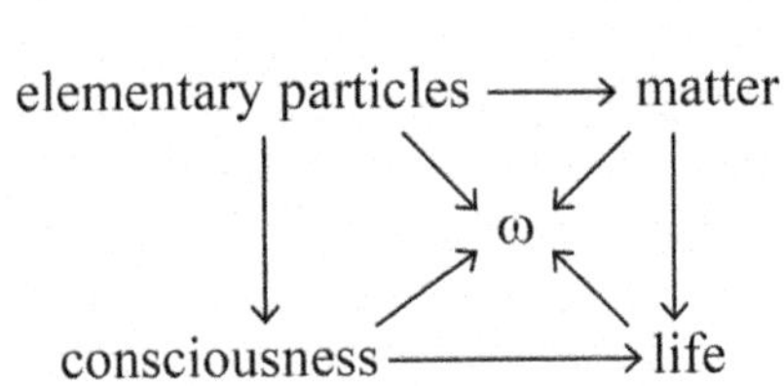

(ω)

and for an initial object α of $S^{\wedge}$,

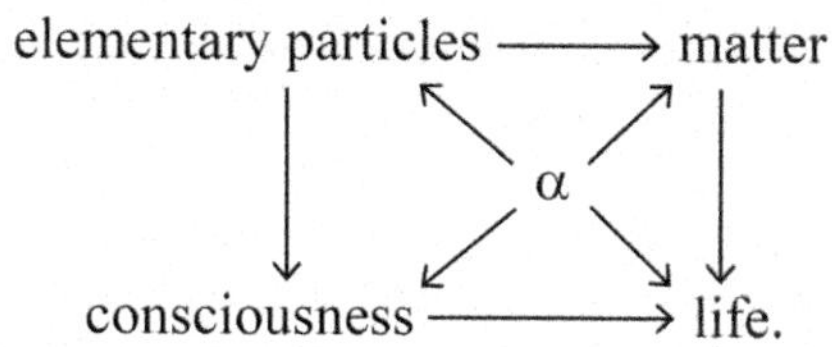

(α)

What is missing in biological sciences and philosophies is the concept of gluing local data to obtain a global object. Namely, we construct a self-organized emerging global object from local objects which are obtained by gluing local data via equalities or isomorphisms over the 'overlaps' of locally given data.

As a macroworld event, let us consider the following example from physics.

If you throw a baseball, e.g., on a baseball field toward the sky with a certain speed v_0, you can determine where the ball will land. In a high school physics course, you learn how to compute where the ball is located after a few seconds, including the exact spot where the ball lands; you can *determine* the position of the ball in terms of the initial speed v_0 and the earth's gravity g and the angle θ from the horizontal level, that is, the location of the thrown ball depends upon these three variables v_0, g, and θ. Namely, when initial conditions in terms of v_0, g, and θ are completely given, you can determine where and how fast the ball is moving. This is classical physics. Note also that the sophisticated version of classical physics is Einstein's theory of general relativity.

Next let us consider a microscopic object, e.g., an electron. A baseball is roughly speaking 10^{12} times bigger than an electron. Having stated the gap in size between them, we now would like to replace the initial conditions for a baseball with an electron. The conclusion is "It does not work." First, it is impossible to determine where an electron is (by the uncertain principle). We must ask why size matters.

However, even a macroscopic object like a baseball consists of elementary particles. The next question to be asked is:

Even though an individual particle's position and speed are fuzzy, for example, a baseball, as an immerged macroscopic object from microscopic objects, the location and the speed of the ball are determined. Why is it possible?

This is the decoherence problem concerning the local issue with the global issue, which should be (and is) relevant to the measurement paradox and the notion of superposition. We will give a temporal topos theoretic interpretation connecting the concepts of decoherence, measurement, and superposition.

Before definitions are given, allow us at this stage to state our main idea prematurely but precisely. In Hermann Weyl's quotation, (%) and (#) parts at the beginning of *Introduction* correspond to the two halves of the following fundamental commutative diagram:

$$\begin{array}{ccc} \alpha & \rightarrow & \omega \\ & \searrow \quad \nearrow & \\ & q & \end{array} \qquad (*)$$

where α is an *initial object*, and ω is a *terminal object* of a *temporal topos* S^ (to be explained in the following chapters).

The right-half (%) of diagram (*) corresponds to our physical existence (phenomena), and the left-half (#) of diagram (*) corresponds to our metal existence (noumenon in the sense of Immanuel Kant). Namely, our interpretation of the *phenomenal world* and the *mental world* are expressed in diagram (*). To give a complete interpretation of the meaning of the commutativity of diagram (*), we begin a long journey in the following chapters.

Suppose that you are sitting still. Then you can suddenly lift your arm. Namely, your desire or thought to lift your arm can be carried into a physical action in space and time. Another example of a phenomenon is, when you think, your brain cells are organizing themselves making your thought reified. We mean that the necessary muscles do not voluntarily move so that it becomes possible for us to think. This is an example of physical entities like brain cells or muscles being initiated and controlled by the mind. Namely, Mind precedes Matter, which is, in our language, expressed as a uniquely determined morphism δ^q from α to an entity q in temporal topos S^. Note that mind and thought are also controlled and organized by physical brain activities. Therefore, there should be a question:

Where does the mind originate? (#)

Our answer to question (#) is it originates from α. As we will explain in the following chapters, diagram (*) plays a crucial role for the interplay between mind and matter. A single-celled organism like bacteria can move; additionally, it is also capable of making choices in its movements. Where does the pseudo-mind of such a single celled living creature like slime mold originate? In terms of temporal topos,

I think; therefore, I am.

is phrased as

The unique morphism δ^q *exists; therefore,* q *must exist.*

As one of the most iconoclastic physicists of the twentieth century, Albert Einstein may come to many readers' minds. Then, who are the most iconoclastic mathematicians? Many of us in the mathematics community consider Alexandre Grothendieck as one of the most iconoclastic and influential mathematicians of the twentieth century. Einstein and Grothendieck both seem to have characteristics of a genius: a person who can hit targets which are invisible to most of us.

Note also that Richard Feynman gave us a warning by saying that some of us should try completely different paths from the popularly taken paths in order to increase diversity in thought and effort; most likely, people will fail, and consequently, no one will remember their names.

Those iconoclastic people may be enjoying the process of attempting such completely different paths.

Our approach has the simplicity of the assumptions. (1) we assume that every entity in the physical universe is represented by an ur-presheaf as an object of temporal topos $S^{\wedge}$. (2) we assume the existence of an initial object α and a terminal object ω in a temporal topos $S^{\wedge}$. Notice that such an initial object α can play the role of the source of mind and life. On the other hand, a terminal object ω can play a role of spacetime. A temporal site, abbreviated as t-site, over which a temporal topos is defined, plays a more fundamental role of the usual time concept, i.e., usual time depends upon the temporal site.

1.3 Category

Responding to the motto of *Springer Nature*:

> *The unique aspect of the motto of integrated science is to cover all aspects of science as one true nature,*

One of the main themes of this treatise is to provide a unified mathematical foundation of natural philosophies. Furthermore, the mathematics which is used for our mathematics of philosophies comes from the notions of categories and sheaves.

Since we have already used several mathematical terminologies and notions, we will provide a brief description for the notions of

> *categories and functors*

in a concise way. See Appendix I for more information on category theory.

A category C is a collection of *things* and the *relations* among *things*. The '*things*' in a category C are said to be *objects*. When X is an object of category C, we write it as

$$X \in Ob(\mathrm{C}). \tag{1.0}$$

Next, the relationship between objects X and Y in category C is expressed as an arrow like:

$$f : X \to Y, \tag{1.1}$$

or

$$X \xrightarrow{f} Y, \tag{1.2}$$

which is expressed as

$$f \in Hom_{\mathrm{C}}(X, Y). \tag{1.3}$$

Such a relationship f is called a *morphism* from X to Y, which is interpreted as an *information flow* from X to Y or sometimes as an *observation* of X by Y. That is, morphism f is an element of the set Hom_C (X, Y) of all the morphisms from X to Y in category C.

For example, when X is measured by Y, object Y receives information on X through morphism $X \rightarrow Y$. For an information flow from X to Y followed by another information flow from Y to Z, we can ask how to phrase the information flow from X to Z. In mathematics, we use the *composition* of two morphisms as follows. Namely, for $f: X \rightarrow Y$ and $g: Y \rightarrow Z$, we define the *composition* of $g \circ f$ as a morphism from X to Z. Namely, we express the information flow from X to Z as the composition of two morphisms f and g, i.e.,

$$g \circ f : X \rightarrow Z. \tag{1.4}$$

After defining the notion of a category, we are ready to explain the notion of a *functor* as follows.

A *functor* is an assignment from a (domain) category to another (codomain or target) category. Let us prepare two categories which we call C and K. Now we are ready to describe a functor F from domain category C to codomain category K. A functor F is also written as an arrow as

$$F : \mathrm{C} \rightarrow \mathrm{K}, \tag{1.5}$$

or

$$C \xrightarrow{F} K, \tag{1.6}$$

just as for a morphism from object X to another object Y as in (1.2).

For any object X in domain C, i.e., $X \in Ob(\mathrm{C})$, functor F assigns an object $F(X)$ in the codomain category K. Then, for a morphism f from X to Y in C, i.e.,

$$f : X \rightarrow Y \text{in C},$$

functor F assigns a morphism $F(f)$ from $F(Y)$ to $F(X)$, i.e.,

$$F(f) : F(Y) \rightarrow F(X) \tag{1.7}$$

in K. That is, F takes objects and morphisms of domain category C to objects and morphisms in codomain category K. Notice that the direction of the arrow in C from X to Y is changed by F to the opposite direction in K, i.e., from $F(Y)$ to $F(X)$ as seen in (1.8). Such a direction reversing functor is said to be a *contravariant functor*. When a functor does not change direction, such a functor is said to be a *covariant functor*. In our temporal topos theory, we use contravariant functors. Note that a *presheaf* is, by definition, a contravariant functor.

As a part of the definition of a functor, a few more additional axioms not described in the Introduction must be satisfied. See Appendix I for additional axioms for a functor to satisfy.

Note Even though categorical and sheaf theoretic notions are fundamental to our theory, we use a more restricted concept of a functor which should be called an *ur-functor* or a *quasi-functor*. Namely, let p be the representing ur-functor for a particle P. Then $p(V)$ is defined only when p is in an *ur-particle state* where V is an object of a t-site. Sometimes we call such a state a *particle ur-state*. Namely, an ur-functor F is only defined for certain objects in a domain category.

We may sometimes call such a restricted ur-functor *simply* a *functor* in the following chapters of the treatise. In our temporal topos theory, the domain category C is replaced by a site S, where a site is, by definition, a category S with a *Grothendieck topology*. See Appendix II for the definition of a Grothendieck topology. Then the definition of a temporal topos S^ is the category of contravariant ur-functors, i.e., *ur-presheaves*, from the domain category called a temporal site S to the *codomain* (or *target*) *category* O which has an initial object and a terminal object. The codomain category O is also said to be an *observation category*.

If a reader has taken a course in category theory or homological algebra, then he or she will recognize the following notation used for the category S^ as the category of *contravariant ur-functors* (i.e., *ur-presheaves*) from S to the codomain category O.

$$S^{\wedge} = O^{S^{opp}}, \tag{1.8}$$

where S^{opp} is the opposite (dual) category of S. The opposite category S^{opp} is defined as follows: objects of S^{opp} are the same as objects of S, but morphism in S

$$f : X \to Y$$

becomes in S^{opp}

$$f^{opp} : Y \to X.$$

Now we conclude our brief description of the notions of a category and a functor.

Let us explain why we use a contravariant functor rather than a covariant functor. Let $p(V)$ and $p(W)$ be the ur-particle states of p over V and W, respectively, where $V \to W$ is temporally linear. The ur-particle state $p(V)$ temporally precedes the ur-particle state $p(W)$. When state $p(V)$ is observed by $q(U)$, where $V \to U$ is a temporally linear morphism in S, $q(U)$ receives an information of state $p(W)$ through the composed morphism $p(W) \to p(V) \to q(U)$. Note that both W and U belong to the ur-lightcone with vertex V. We will come back to this topic on special relativity and ur-lightcones in Chap. 3.

1.4 Our Approach

Phenomena of emerging objects, i.e., global entities, as self-organizing discretely given individuals in nature, often appear to be mysterious and puzzling. We are surrounded by these phenomena. For example, trees and animals consist of cells, which consist of atoms, which, furthermore, are compounds of elementary particles. When we observe something emerging from small entities to form a large entity, we wonder about the mechanism behind the self-organization of small and trivial entities to be able to form an organized large entity. Such global objects need not be physical matter. For example, when one listens to one individually trivial musical note, one may not have any idea about a global non-trivial well-known music piece. However, when one continuously listens to enough notes, a global music piece emerges and becomes recognizable. For this example, we first need to mathematically formulate the mental (or psychological) notion of understanding or recognition before formulating how to glue discretely given local data.

We will provide a structurally common mechanism for the physical and cognitive phenomena in terms of categories and sheaves. Namely, we will specify the underlying mathematical common structures in terms of sheaves and categories for these phenomena.

The procedure of a reconstruction of a global object from locally trivial data is the method of descent theory; it is like constructing a world map by gluing small pieces of local maps. Once such a global object is obtained by '*gluing*' the locally given data, then we can *descend* the global object toward the local data by restricting the global object. Then these restrictions of a global object must match with the original local data.

Hence, as the first step, our task is to specify the mathematical expressions for '*gluing*' local data and '*restricting*' the global entity to each local datum.

In what will follow, we will specify by *for mathematically inclined readers* to indicate that such technical descriptions and formulations are not meant for people from a humanities background.

We can consider a brain and its functioning ability which emerges from trivial objects like individual cells and ultimately elementary particles. We emphasize that explaining these procedures from the local to the global is the theme of this treatise.

In the history of sciences, when studying a certain living or non-living object, as a first approach, we have analyzed an object by decomposing the object into smaller objects, and if needed, further decomposition may be carried out. In physics, this is a typical method of research activity; we have been trying to reach such ultimately fundamental entities like elementary particles or even strings as building blocks. However, our theme is the other direction, i.e., from the small to the large. In mathematics, a procedure from the local (i.e., the small) to the global (i.e., the large), by gluing locally given data, is said to be the method of sheaf or descent theory.

Our emphasis is "one true nature." Our methods for this goal are provided by categorical and sheaf theoretic (and sometimes stack theoretic) approaches.

Any unifying physics theory describing simultaneously the microcosm and macrocosm deserving to be called a quantum gravity theory, must be consistent with classical physics of Newton and Einstein and furthermore with quantum physics. Yet, we have the other half of the reality, i.e., that of the cognitive or mental world, the left-half of diagram (*).

In this treatise we will provide needed background for the *temporal topos* theory, often abbreviated as *t-topos* theory, to be applied to both physical and cognitive phenomena. Even if the right-half of diagram (*), i.e., the physical aspect of reality, is unified, any theory should not be regarded as a theory of 'everything' as a description of the ultimate reality until one can first give an explicit definition of being animate or inanimate and can explain explicitly the cognitive states of a conscious entity including self-consciousness. The phenomenon of Organization-Emergence in life-sciences must be addressed as a process from the local (the small) to the global (the large).

For any entity (particle), assign an '*avatar-object*,' so to speak, in a category called a temporal topos S^. An avatar-object of that category S^ is said to be an *ur-presheaf* (which is a contravariant ur-functor as explained earlier) in the temporal topos S^. That is, for any entity Q, there corresponds an ur-presheaf q in S^, as an avatar of Q. One of the reasons we replace a physical object Q with the corresponding ur-presheaf (avatar-object) q is that in category S^ we can express a flow of information between objects as a morphism between them. Notice that a morphism does not exist in the physical world!

Let p be the ur-presheaf representing physical entity P. That is, p is the representing ur-presheaf for P, i.e., the avatar of entity P. When entity P is observed by entity Q, there is a morphism f from p to q in the category S^. Then we write it as

$$f : p \to q,$$

meaning q receives information from p through f.

To formulate the concepts in terms of categories and sheaves, it is crucial to express and emphasize the relationships in terms of morphisms, i.e., arrows, not in terms of objects. Namely, important concepts should be expressed as properties of morphisms as the relationships among objects. That is,

morphisms are more crucial

than objects themselves. This is the philosophy and the method of Alexandre Grothendieck for his fundamental philosophy of his mathematics found in his study of algebraic geometry. In our case, importance is placed on a morphism between ur-presheaves, namely, *for mathematically inclined readers*, i.e., technically speaking, a morphism between presheaves is a *natural transformation*, or simply a *morphism of contravariant functors*. See Appendix I, [Gelfant-Manin (1996)], or [Kato (2006)] for related topics and such mathematical treatments as these.

As we mentioned earlier, in modern scientific fields like physics, chemistry, and biological-life sciences, the typical direction of study is to decompose and analyze a larger object via the consisting small entities. That is, we have learned more fundamental laws from the smaller components, which is typically considered the modern scientific approach in physics, chemistry, and biology, to understand the emerged global entities. The other direction of building a global object from the local entities, e.g., emergence through self-organization, is usually more difficult, which is often referred to as the phenomena of Organization-Emergence.

In the fields of pure mathematics, e.g., in the theory of holomorphic functions of several complex variables, the concept of coherence was crucial for discovering the notion of a sheaf, which was applied to solve the *Cousin problems* during 1930s. See, e.g., H. Grauert's superb article in *Miscellanea Mathematica,* Springer-Verlag, 1991. However, our situation is easier since we are already given an emerged object. Namely,

> *we already know what the global object is. Our task is to define the notion of 'how to glue local data' and the notion of 'restriction' so that a global object can descend properly to local data, i.e., matching with local data.*

Using an example from physical sciences, we can demonstrate what we mean by a process from local data to a global object.

An electron in a stone and an electron in an apple are indistinguishable as electrons. But we distinguish one from the other in the sense that they are represented as isomorphic ur-presheaves. We express this situation as they are isomorphic as ur-presheaves which can have different states, e.g., one electron is in an ur-particle state and the other electron is in an ur-wave state.

We ask two questions:

> *Why does knowing the individual parts not tell us the whole which consists of those parts?*

and

> *How do we phrase the above phenomenon in terms of morphisms?*

Such a phenomenon should be expressed as the *non-composability of morphisms* which means two arrows in the *opposite directions.* Let us be more explicit to phrase the non-composability of morphisms as follows. The meaning of the non-composability implies the following consequence: we cannot guess a global object like an apple by observing an individual component entity like an electron, an atom, or a cell in the apple.

1.5 Parts and a Whole

To phrase this issue on a solid foundation, we must be a little more technical in terms of mathematical concepts *for mathematically inclined readers.*

Let A be the representing presheaf for an apple A. Then decomposing the apple into small entities (like cells, molecules, etc.) corresponds to expressing A as a *co-product* (or *direct sum*) of sub-presheaves $\{e_j\}$ as

$$A = \coprod e_j, \tag{1.9}$$

where each component e_j of the co-product $\coprod e_j$ is the representing *ur-subpresheaf* corresponding to, e.g., an electron or a molecule in the apple. There is a property of a co-product in mathematics, which says there exists morphism ε_j: $e_j \to \coprod e_j$ for each j, which could be interpreted as e_j being a part of the whole $\coprod e_j$. For the mathematical background see Appendix I. Note also that the global object consisting of $\{e_j\}$ as local given data is presheaf A, i.e., there is no need to *construct* a global object from $\{e_j\}$. That is, our task is to define and formulate how to '*glue*,' via isomorphisms, local information $\{e_j\}$ and how to '*restrict*' the global object A to each local e_j.

When a smaller component e_j of the whole entity $\coprod e_j$ is observed by an observer q, we have an observation morphism f: $e_j \to q$, which is an information flow from e_j to q as noted earlier.

We are now ready to explain the meaning of the mathematical expression of the *non-composability*: explaining why observing a smaller trivial part tells nothing about the global whole. It is similar to the idea of seeing only a strand of hair on a tail of an animal does not tell us what the animal is.

See the following diagram:

$$\begin{array}{ccc} \coprod e_j & \leftarrow & e_j \\ & & \downarrow \\ & & q \end{array} \tag{1.10}$$

where, as noted earlier, $\coprod e_j$ is a co-product consisting of individual local ur-presheaf objects $\{e_j\}$. We now can mathematically explain that observing an individual e_j by observer q does not provide information of the large $\coprod e_j$.

Recall: for morphism $\xrightarrow{f}$ followed by morphism $\xrightarrow{g}$ in the same direction, morphism f can be composed with g as the morphism $g \circ f$. However, morphism $\xleftarrow{h}$ in the opposite direction of morphism $\xrightarrow{g}$ cannot be composed, which is the situation in the above case.

Namely, we have morphism $\coprod e_j \leftarrow e_j$ which is the opposite direction of morphism $e_j \to q$. Therefore, we cannot compose these morphisms, which implies that there does *not* exist a morphism from the large $\coprod e_j$ to an observer q in diagram (1.10), indicating the impossibility of receiving any information from the global object $\coprod e_j$, a *big picture*, so to speak, by just observing the small part e_j.

Namely, there does not exist a morphism from a global object $\coprod e_j$, e.g., an apple, to observer q even when there is a morphism from an individual e_j to the observer q. Notice also that when a global object $\coprod e_j$ is observed by q, which means that there is a morphism ϕ (a measurement or observation morphism) from $\coprod e_j$ to q, then the restriction of the global object $\coprod e_j$ to an individual e_j is nothing but the composition $\phi \circ \varepsilon$ from e_j to q of the canonical morphism ε with ϕ. See the following diagram:

$$\coprod e_j \leftarrow e_j \quad \searrow \quad q. \tag{1.11}$$

In mathematics, there is a notion of a restriction ε_j^* for $\coprod e_j \leftarrow e_j$ of a global object to each datum.

The next question to be asked is:

In what way, in terms of presheaves, are such elementary particles organized to emerge as a global non-trivial object like a stone or an apple?

This opposite direction of decomposition, that is, from the small to the large can be rephrased as follows.

What is the mechanism and formulation to go from a given local data by gluing (i.e., organizing) the locally given trivial entities (e.g., particles, cells, or even individual ants), to form a global object e.g., an apple, a brain, a colony of ants, or even a conscious human?

As we will show in detail, we prepare two categories S and S^. We call such a category S, whose objects play parameters, a *temporal site*, abbreviated as a *t-site*, and S^ a *temporal topos*, abbreviated as a *t-topos* satisfying the temporal topos theoretic constraints. See the following chapter for these constraints.

We would like to show that only the existence hypotheses of an initial object α and a terminal object ω of a temporal topos S^ can provide the needed structures for unifying phenomena simultaneously in cognitive and physical sciences.

Hence, our treatise's task is to convince the readers how and why the sheaf theoretic and categorial role played by α as an initial object of t-topos is suited for describing cognitive phenomena, and the sheaf theoretic categorical role played by ω as a terminal object of t-topos is suited for physical phenomena.

Having said that, from the very beginning we could call such an initial object α of temporal topos S^ an *omnipresent and universal cognitive source*; it might be smoother *for a more philosophically inclined reader* to see why and how various concepts in the t-topos theory are introduced. However, we prefer to show that the mathematical consequences obtained by α, being an initial object of t-topos S^, indicate such a philosophical and historical entity. Namely, the name of an initial object α does not matter, but the properties and roles obtained by such an initial object do. Note that *omnipresent-ness* is captured and formulated as an initial object α being a constant functor, i.e., independent of generalized time periods, i.e., objects of the t-site.

Our second fundamental principle is as follows:

Wherever emergence and self-organization occur in nature, 'gluing' via *isomorphisms on 'overlaps' of local data should be applied.*

We wrote 'gluing' and 'overlaps' since we will define those concepts precisely in terms of sheaves.

The main topic of this treatise is to provide a mathematical formulation for the notion of *Gestalt*, which is understood as an organized whole consisting of its trivial parts.

As example for the phenomena from individual entities to an organized-emerged unified entity, we can consider the relationship between individual ants $\{a_i\}$ and an ant colony $A = \coprod a_i$. Each ant's behavior is simpler and more trivial, i.e., not organized as a whole; however, the ant-colony as an organized whole is complex and nontrivial.

As another example, we can also consider a series of musical notes as local data, which emerge as a music piece as a global object, namely, the transition from each discrete piece of information from locally given data to global information. A much more challenging example might be the process from a DNA sequence to the determination of a global form, crucial for the study of embryology. We will also formulate the notion of AI, artificial intelligence, as mimicking a human brain and consciousness itself.

All these examples involve the process of the local to the global, or the small to the large.

Our categorical and sheaf theoretic methods provide *only structural aspects* of phenomena. Having said that, as we will explain in Chap. 3 on the applications of t-topos to quantum physics, one can associate a micro-morphism in a *t-site* with Planck-scale for a particle p in the following way: for a micromorphism $V \to W$, assign the usual Planck time to the difference of two states from $p\ (W)$ to $p\ (V)$. For physics related papers as applications of temporal topos, see e.g., [Kato (2004, 2005, 2010, 2011, 2013)], where $\omega = (\tau, \kappa)$, a final object of $S^\wedge$, plays the role of spacetime in quantum and relativistic aspects.

The so-called theory of '*everything*' in physics was supposed to be a theory that unites all the fundamental forces in nature. Yet, we can ask:

> *What is the mechanism for small entities, e.g., particles, being gathered and organized together emerging as, e.g., a functioning human brain capable of creating mathematical thoughts, literature, and art?*

On the other hand, sometimes such gathered individually trivial small entities become just a stone. Symbolically speaking, such a theory of everything is *missing the other half of the universe*. See diagram (*).

Let us quote Einstein's words:

> "*You imagine that I look back on my life's work with calm satisfaction. But from nearby it looks quite different. There isn't a single concept of which I am convinced that it will stand up, and I feel unsure if I am even on the right road. My contemporaries, however, see me as a heretic and reactionary who has, as it were, outlived himself.*"
>
> From a letter written by Einstein to Maurice Solovine in 1949, p. 96, *Einstein – Decoding the Universe*, by Françoise Balibar, Discoveries, Harry N. Abrams, INC., Publishers, 2001.

After extremely successful descriptions of physical aspects in numerical accuracy and predictability of the macrocosm and the microcosm by general relativity and quantum theory, respectively, we await the establishment of one theory for both macrocosm and microcosm, which is often called *quantum gravity*. Namely, the task is to bring the small and the large together under one umbrella, so to speak.

On the other hand, what about the phenomena of the living? We can wonder how such sequences of four proteins inside the nucleus of a cell can matter at all to affect globally the shapes and colors of animate entities. Such an effect is more puzzling than, for example, in the seventeenth century, how the earth can possibly affect the movement of the moon which is 384,400 km away from the earth. It may be equally puzzling to understand how a colony of ants (e.g., a leaf-cutting ant colony) can organize individual ants without the use of language. What is the mechanism behind such a behavior connecting local individual triviality to the global non-triviality?

One of the main themes of this treatise is to initiate such a method in terms of sheaf theory to investigate how individually trivial entities are organized and emerge as a highly non-trivial global object.

It seems to be an appropriate time for *mind and matter* to be gathered under one *roof* through mathematical formulations. The shape of such a *roof* in terms of a commutative diagram may look like the following with the roof consisting of an initial object α, playing the role of *mind*, and a terminal object ω, playing the role of *matter* for all the entities q in a temporal topos $S^{\wedge}$:

$$\begin{array}{ccc} \alpha & \rightarrow & \omega \\ & \searrow \ \nearrow & \\ & q. & \end{array} \qquad (*)$$

For example, q can be an ur-presheaf representing an electron or a conscious entity like a human being. This commutative diagram in $S^{\wedge}$ is the most fundamental diagram in our temporal topos theory as a unifying view of q for the cognitive aspect $\alpha \rightarrow q$, and the physical aspect $q \rightarrow \omega$. Furthermore, diagram (*) represents our epistemology and ontology as we will explain in the following chapters.

Notice that those initial and terminal objects are constant functors in the functor category $S^{\wedge}$. That is, we have $\alpha(V) =$ (or isomorphic to) $\alpha(W)$ for any V and W in t-site S. In this sense, objects α and ω have '*absolute*' and '*omni*' existence. Namely, as constant functors the 'values' of those functors do not vary for varying objects (generalized time periods) in a site S. Newton might have intuited that the fixed (absolute) nature of the background space–time should be taken as a constant functor ω with respect to generalized time periods, i.e., objects of t-site S. The usual time concept τ is invariant with respect to objects in t-site. However, we expediently write $\tau(V)$ when it is convenient for explanations for special relativity effects.

We can build a theory without usual time; instead, a t-site plays the role of *true* time as generalized time. In that sense, the temporal topos approach is relevant to Julian B. Barbour's theory. See [Barbour (1999)] and [Craig Callender and Nick Huggett, editors (2001)].

Many commutative diagrams will appear in the following chapters. The commutativity of a diagram may be interpreted as the consistency, comparability, and coherency for various situations. Namely, for anything to exist at all, the commutativity of diagram (*) must be satisfied, viz., for q to exist we must have an agreement between the physical and the cognitive, in other words, the commutativity of diagram (*):

$$\zeta = \sigma_q \circ \delta^q : \alpha \to \omega.$$

Lastly, we do not hesitate to recall and review definitions and notions having appeared in earlier chapters for the sake of continuity and smoothness, which should be helpful to avoid interruption of the reader's thought.

1.6 Synopsis

We describe our themes and methods in this introductory chapter. Our main theme is to show how to organize a collection of small trivial objects like atoms and cells to build an emerged non-trivial global object like a living creature.

Knowing the present state does not tell us the past state. That is, for example, if we see a (Schrödinger's) dead cat, we cannot tell what caused the death of the cat. However, if the present state is known, we have some idea what a future state may be. In classical physics, when a complete description of the initial conditions is given, the future state is determined. But in the quantum world, even a position cannot be measured with perfect accuracy because of the uncertainty principle.

We can formulate the knowledge of the future via an observation of the present state more precisely as follows. We can explain why, for our temporal topos theory, the usage of a contravariant functor rather than a covariant functor is appropriate. Let V be an object of a temporal site S corresponding to the present state of object q in a temporal topos $S^\wedge$. When q observes p over V and W, respectively, we have ${s^W}_V$: $p(W) \to q(V)$. Then $f{:}W \to V$ is necessarily a temporally linear morphism. The future ur-state of p over $W*$, i.e., $p(W*)$, may be predicted in the sense that we have the canonically induced morphism $p(W) \leftarrow p(W*)$ by contravariant ur-functor p. By composing observation morphism ${s^W}_V{:}\, p(W) \to q(V)$ and canonically induced morphism $p(f)$: $p(W*) \to p(W)$ in diagram.

$$\begin{array}{ccc} p(W) & \leftarrow & p(W^*) \\ \downarrow & \swarrow & \\ q(V), & & \end{array}$$

we have an information morphism $p(W*) \to q(V)$. That is, the composed morphism

$$s_V^W \circ p(f) : p(W*) \to q(V)$$

provides an information about the future state $p(W*)$ of p over $W*$ through the observation of the present state $p(W)$.

On the other hand, if $W^{\#}$ is an object of S corresponding to a past state of p as $p(W^{\#})$, i.e., a temporally linear morphism g: $W^{\#} \to W$, we have the induced canonical morphism $p(g)$ from $p(W)$ to $p(W^{\#})$ as

$$p(W^{\#}) \leftarrow p(W).$$

An observation of the present state $p(W)$ by $q(V)$, i.e., $s^{W}{}_{V}: p(W) \to q(V)$ as before, cannot be composed with $p(g)$. See the diagram below.

$$\begin{array}{ccc} p(W^{\#}) & \leftarrow & p(W) \\ & & \downarrow \\ & & q(V). \end{array}$$

1.7 Glossary

Notes

(i) We use the words *world*, *universe*, and *cosmos* interchangeably. However, we in general have the inclusions as follows:

the world $\subset$ the universe $\subset$ the cosmos.

(ii) We use cognitive, mental, and sentient as synonymatic words to describe the left-half of the fundamental commutative diagram (*).

R is the set (field) of real numbers. For example, n-product of **R**, i.e., $\mathbf{R} \times \mathbf{R} \times \cdots \times \mathbf{R} = \mathbf{R}^n$.

C is the set (field) of complex numbers. We have also $\mathbf{C}^n = \mathbf{C} \times \mathbf{C} \times \cdots \times \mathbf{C}$.

We use letters A, B, … as objects of a category, and q, p, … are often used for objects in category $S^{\wedge}$. On the other hand, letters C, D, K, … are used for categories.

F: C $\to$ D indicates a functor from category C to category D. We sometimes denote as F(-) or $F(\bullet)$ for an unspecified object in D when an object in C is not determined. As an example, for a presheaf q we write q(-) or $q(\bullet)$ when q is not reified at an object in site S.

$f: A \to B$ indicates a morphism from object A to object B in a category.

$g \circ f$ is a composed morphism of f and g. See also Appendix I.

For morphisms f: X $\to$ Y, g: Y $\to$ Z and h: X $\to$ Z, the following diagram is said to be commutative when $h = g \circ f$ holds.

$$\begin{array}{ccc} X & \rightarrow & Y \\ & \searrow \; \swarrow & \\ & Z & \end{array}$$

$\{e_j\}$ or $\{e_j\}_j$ indicates a family or a set of $e_1, e_2, \ldots$, where index j belongs to an index set J, which is written as $j \in J$.

$\coprod$ indicates the coproduct or direct sum of objects in a category.

$Hom_C(A, B)$ indicates the set of all morphisms from object A to object B in category C.

$A \in Ob(C)$ means A is an object of category C.

$k_q{}^V{}_W$ or $j_q{}^V{}_W$ is a non-canonical morphism (sometimes also as $k^V{}_W$) from $q(V)$ to $q(W)$, written as $k_q{}^V{}_W = k^V{}_W: q(V) \to q(W)$.

$V, U, W, \ldots$ are used for objects of a site S, which are also said to be generalized-time periods.

$\omega = (\tau, \kappa)$ is a (spacetime) terminal object in temporal topos $S^\wedge$.

σ_q is a uniquely determined morphism from any object q in $S^\wedge$ to a terminal object ω in $S^\wedge$.

δ^q is a uniquely determined morphism from an initial object α to any object q in $S^\wedge$.

$s^V{}_U$ is an observation or measurement morphism which is written as $p(V) \to q(U)$. Morphism $p(V) \to q(U)$ also indicates that an ur-particle state of p over V is observed or measured by q over U, where $V, U \in Ob(C)$. Note that in the case where $p = q$, an observation morphism is a mnemonic morphism.

S is a temporal site (abbreviated as a t-site).

O indicates codomain category with an initial object A_O and a terminal object Z_O.

$S^\wedge$ is a temporal topos (abbreviated as a t-topos).

Z is the set of integers $\{\ldots, -3, -2, -1, 0, 1, 2, \ldots\}$.

$\mathbf{Z}^+$ is the set of positive integers $\{1, 2, 3, 4, \ldots\}$.

Chapter 2
From Local Data to Global Entity

Contents

> *I have never forgotten how, during my next to last year in school, I happened in the attic of my parent's home to come upon a copy of a short commentary on Kant's Critique of Pure Reason, dating from 1790, yellowed and foxed by long storage. Kant's teaching on the "ideality of space and time" immediately took powerful hold of me; with a jolt I was awakened from my "dogmatic slumber" and the mind of the body found the world being questioned in radical fashion.*
>
> *Mind and Nature, selected Writings on Philosophy, Mathematics, and Physics*, by Hermann Weyl, Chap. 9 'Insight and Reflection,' p. 205.

In Chap. 1, we described our basic themes, and we also provided a brief introduction to our categorical methods. In this chapter, we will focus on further topics and examples relevant to our themes, and the focus will be more technically mathematical descriptions of temporal topos methods. If you are more interested in the humanities fields rather than mathematical sciences, you can read quickly though the portions of this chapter written *for mathematically inclined readers*.

We are surrounded by physical and cognitive phenomena. As we have already mentioned, some of those phenomena have puzzling and curious issues:

1. *In biology*: a collection of, e.g., atoms and cells can end up with a conscious living creature like a human being who can think. Is there an explicit mathematical formulation to explain the process from the collection of trivial entities to an organized large entity?

G. C. Kato, *Temporal Topos Methods for the Philosophy of Natural Sciences*,
https://doi.org/10.1007/978-981-96-2420-1_2

2. *In cognitive sciences*: we have a mind and body. Our *will* in the mental domain, e.g., the desire to move our fingers, can be reified and expressed as a physical action of e.g., playing the piano. How can such a *mentally initiated will* be expressed as physical actions?
3. *In embryology and biology*: a change in DNA sequences can affect a change in our body. How does it work? What is the procedure for it?
4. *In physics*: what (‘*pseudo-*’) *was* going on before the Big Bang? Formulate such a state of the universe. We are assuming such a grand beginning existed at the birth of our physical universe.
5. *In philosophy*: does there exist a source-like entity for life and mind which is, in an organized way, affecting all the entities in the universe consistent with physical laws?

These are naïve (yet profound) questions we might have asked ourselves during our childhood or youth.

We look for a common language to formulate phenomena in these fields. Especially in questions 1, 2, and 3 in the above list, we need to ask what the mathematical formulations are for the processes from the locally trivial objects to a large non-trivial object. The mathematical language we use comes from category theory, especially sheaf theory.

Locally trivial and isolated objects can be cells, molecules, atoms, and ultimately elementary particles like an electron and a photon. We say non-trivial global objects as emerged objects organized from small objects, can be fungi, trees, animals, and ultimately a human being. For the cognitive aspects of the interplay between the local and the global, see Chap. 4, Cognitive and Biological Sciences.

As preparation for our language of a temporal topos, we need to provide some constraints on the usual concepts in category theory and sheaf theory. A topos is the category of presheaves over a site. A site is a category with a Grothendieck topology. See Appendices I and II for the definitions of a category and a sheaf.

As we have already mentioned in the brief description in Chap. 1, by definition, a presheaf is a contravariant functor from a category to a category. For our temporal topos theory, we need a more restricted notion than a presheaf. Such a more restricted presheaf is called an *ur-presheaf*. Hence, we must note:

an ur-presheaf is not a presheaf.

Consequently, we need a new name for a functor. We can call such a restricted functor as either an *ur-functor*, a *quasi-functor*, or a *pseudo-functor*. However, ‘*ur*’ (meaning ‘primeval’or ‘primitive’) is our choice for this treatise.

Before we explicitly describe a set of axioms for a topos to be a temporal topos, we shall first briefly review our principal approach.

For any entity Q in a physical realm, we assign an ur-presheaf q, which may be considered as the ‘avatar’ of entity Q. Then we say that q is the *ur-presheaf associated with* Q. As we mentioned in the Introduction, q is called the *corresponding ur-presheaf to entity* Q. Also, recall that q is an ur-functor from a domain category S, called a temporal site, abbreviated as a t-site to a codomain category O. Note that we assume domain category O has an initial object and a terminal object. An initial object α is an object in a category C satisfying the following condition: for any object q in C, there exists always a unique morphism δ^q from α to q. If we use the notation in Chap. 1, the uniqueness can be expressed as

$$Hom_{\mathrm{C}}(\alpha, q) = \{\delta^q\},$$

for an arbitrary q in S^. Equation $Hom_{\mathrm{C}}(\alpha, q) = \{\delta^q\}$ indicates that δ^q is *the only* morphism from α to q. We often say that morphism δ^q is the uniquely determined morphism from α to q. Similarly, for a terminal object ω, we have

$$\mathrm{Hom}_{\mathrm{C}}(q, \omega) = \{\sigma^q\}$$

for an arbitrary object q. For our t-topos theory, we replace C with S^.

We write such an ur-presheaf q, i.e., an ur-functor, from a temporal site S to a codomain category O

$$q : \mathrm{S} \to \mathrm{O} \qquad (@)$$

as in the Introduction. Temporal topos S^ is the totality of these ur-functors $\{q\}_{q \in Ob(\mathrm{S}^\wedge)}$ with a few additional structures beyond being just a topos. Having reviewed a few notions from Chap. 1, we are ready to describe the temporal topos theoretic constraints, i.e., axioms, for a topos to become a temporal topos.

We need to provide a more precise description in what will follow *for mathematically inclined readers.*

Category S^ is said to be a *temporal topos*, and a site S is said to be a *temporal site*, which are abbreviated as *t-topos* and *t-site*, respectively, when the following axioms are satisfied.

Before we list the axioms for a temporal topos, there are a few remarks to be made. We denote an initial object as α and a terminal object as ω of temporal topos S^. As will be shown in Chap. 3, note that a terminal object ω of temporal topos S^ is associated with spacetime. Since ur-presheaves α and ω are constant ur-functors, for any object W of a site S, α and ω, assign an initial object A_{O} and a terminal object Z_{O} of codomain category O, respectively. To be explicit: $\alpha(W) = A_{\mathrm{O}}$, and $\omega(W) = Z_{\mathrm{O}}$ for an arbitrary object W in the domain category S. However, we sometimes use $\omega(W)$ for expediency's sake, even though ω is a constant functor with respect to a parameter object W in S.

For an arbitrary object q in a t-topos S^, we have a uniquely determined morphism σ_q from object q to ω which is the definition of a terminal object of a category; similarly, there exists a uniquely determined δ^q from an initial object α to q. Then the following fundamental diagram (2.4).

$$\begin{array}{ccc} \alpha & \rightarrow & \omega \\ & \searrow \quad \nearrow & \\ & q, & \end{array} \qquad (*)$$

Which appeared in the Introduction, commutes. Diagram (*) is also philosophically fundamental for our approach; the existence of any entity, represented by an ur-presheaf q, has two aspects. The physical aspect of q is represented by σ_q as an *embedding into spacetime* ω, and the cognitive aspect of q is represented by δ^q as *receiving information from* α.

For mathematically inclined readers, we have a unique morphism δ^ω from α to ω, and also unique morphism σ_α from ω to α, regarding ω just as an object of S^ and regarding α just as an object of S^, respectively. Then the uniqueness of such a morphism *either* for an initial object *or* a terminal object implies σ_α and δ^ω are the same. We call this morphism ζ for the common morphism from α to ω. That is, we have a uniquely determined morphism

$$\zeta : \alpha \to \omega$$

in (*), which is expressed as $\zeta = \sigma_\alpha = \delta^\omega$.

Remarks

(1) This categorical approach is flexible in the sense that all initial objects are isomorphic, i.e., essentially the same, and all the terminal objects are also isomorphic.
(2) The two objects α and ω are constant functors since α and ω are an initial object and a terminal object of category S^, respectively, as we have noted earlier. However, for example, we write $\omega(W)$ (= a terminal object of the codomain category of S^ for any $W \in Ob(\mathrm{S})$) for expediency's sake even though ω is independent of any object in S^, i.e., ω being constant with respect to $Ob(\mathrm{S}^\wedge)$. It can be useful when considering a temporally linear morphism $V \to W$ of S to indicate W as in the ur-lightcone with vertex V. We will show, in Chap. 3 on physics topics, that an information on p, e.g., the mass of p, measured by q through the spacetime effect caused by p, is independent of the distance (and time) between the location of p, i.e., ur-state $p(V)$, to the space(time) ω.

Brief Historical Note for the Concept of Space The simplest example of space may be a one-dimensional line, denoted as $\mathbf{R} \underset{=}{\text{def}} \mathbf{R}^1$. Then, we can consider a higher dimensional space, viz., an n-dimensional Euclidean $\mathbf{R}^n$, $n > 3$, rather than a naïve two-dimensional Euclidean plane $\mathbf{R}^2$ and three-dimensional Euclidean space $\mathbf{R}^3$. However, in the nineteenth century, a significantly more sophisticated concept of a *manifold* was introduced by Georg Bernhardt Riemann. A high dimensional manifold is locally isomorphic to $\mathbf{R}^n$ or a complex n-dimensional space $\mathbf{C}^n$. A familiar example of a manifold for $n = 2$ would be a torus, i.e. a surface of a doughnut, which locally looks like $\mathbf{R}^2$. As a more abstract space than these *matric spaces* where a notion of distance between points is defined, we can consider a *topological space*. A topological space is a more abstract space than Euclidean spaces in which even a distance concept may not be defined. In mathematics, a space has also been captured as a set where functions are defined. The concept of a *sheaf* comes from a collection of functions defined over such a set. Among these generalizations of concepts of a space, one of the ultimate space concepts is that of a *site* (def *a category with a Grothendieck topology*) generalizing a space concept topologically and dimensionally, and the space of functions defined over the space. See [Artin (1962)] or [Schubert (1972)].

It is said that the physical universe consists of elementary particles or even strings. Pythagoras considered that numbers are ruling the universe. Symbolically speaking,

the universe consists of numbers.

As a symbolical phrase for our methods,

the universe is filled with ur-presheaves.

Even though we will introduce our philosophical views in Chap. 5, our approach toward the physical universe is the following.

As noted, every physical entity Q is associated with ur-presheaf q. That is, we regard the associated ur-presheaf q as the '*avatar-like* object' of a physical entity Q. Some people prefer the reverse view, i.e., the physical entity is the avatar of the ur-presheaf. Each view may depend upon one's ontological view, e.g., pro-idealism versus pro-realism.

Note also that we *do not assume* that for an arbitrary ur-presheaf, there corresponds a physical entity. That is, there are more ur-presheaves than physical entities.

2.1 Temporal Topos Axioms

All the entities existing in the physical universe, regardless of possibility of being observed or not, are going to be represented as ur-presheaves, i.e., as contravariant ur-functors. The totality of all these ur-presheaves is called a temporal topos with the following structures. Then, the domain category for these ur-presheaves is said to be a temporal site. A *temporal topos over a temporal site* differs from a general (usual) topos over a general (usual) site. The required conditions for a topos to be a temporal topos are given by the following axioms.

See [Kato (2004, 2005, 2010, 2013)] for more discussions of relevant materials especially to quantum physics in this chapter.

The very first axiom [*i*] is fundamental for our view of *reality* via categories and sheaves. If we accept Plato's perspective, physical entities are only shadows, and ur-presheaves are real. See Chap. 5, Philosophical Roles of Temporal Topos.

Symbolically speaking, our approach is.

ur-functor over matter.

Now formally, we list the axioms for temporal topos, even though we have already mentioned them earlier.

[i] For any particle Q, e.g., an electron or a cell in either microcosm or macrocosm, there corresponds an object q in a t-topos S^. Then ur-presheaf q is said to

be the *corresponding ur-presheaf to physical particle Q*, or q is the *associated ur-presheaf with particle Q*.

Notes

(1) However, we repeat that for an arbitrary ur-presheaf in temporal topos S^, there may not exist a particle. In other words, there are more ur-presheaf objects in S^ than all the particles in the physical universe.

(2) Chris Isham made a humorous remark on the notion of an ur-presheaf associated with a physical particle; he said "… *even an elephant will do*," during the 2nd International Workshop on *Sheaves and Topoi in Physics* at Imperial College, 17–19/7/2003. See also page 169 in [Mallios and Zafiris (2016)].

Here is another fundamental axiom [*ii*] which tells us that we only consider ur-presheaves $\underline{\underline{\text{def}}}$ contravariant ur-functors in our theory, *not* presheaves $\underline{\underline{\text{def}}}$ contravariant functors.

[ii] Object q in a temporal topos S^ need not be defined for an object of a *temporal site* S. We say that q is *reified at* an object V of t-site S if and only if q is defined over V as $q(V)$. Then q is said to be in an *ur-particle state over V or at V* when q is defined over an object V of a t-site S. Otherwise, an ur-presheaf q is said to be in an *ur-wave state*. Namely, $q(V)$ is in an ur-particle state over $V \in Ob(\mathrm{S})$ if and only if q is reified at object V in a t-site S. We write simply q, or sometimes $q(\text{-})$, when q is not reified, i.e., not defined over any object in S, i.e., when q is in an ur-wave state. That is, we regard an ur-presheaf q itself as a particle representing an object in an ur-wave state.

Remarks

(1) As noted earlier, if q is just a functor, then by the definition of a functor, q is defined at an arbitrary object in a domain category. However, for temporal topos theory, an ur-functor q as an object of temporal topos S^ is defined at an object of a site S *only when* q is in an ur-particle state, namely, the corresponding entity Q is in a particle state. We repeat that an object of a temporal topos is an ur-functor, i.e. *not* a functor in the usual sense.

Hence, it is important to be aware that an ur-functor is not a functor, an ur-presheaf is not a presheaf, a temporal topos is not a topos, and a temporal site is not a site.

That is, for any object q in t-topos S^, there are exactly two cases:

Case 1: ur-presheaf q is in an ur-wave state, i.e., q is not reified at an object of S.

Case 2: ur-presheaf q is in an ur-particle state, i.e., q is reified at an object V in S as $q(V)$.

The *wave-particle duality* of a particle Q is interpreted as whether the corresponding ur-presheaf q in a temporal topos S^ is reified or not.

(2) For the fundamental commutative diagram (*), which first appeared in the Introduction, the notion of a *milieu* of q is defined as follows. The triple ($\delta^q(V)$, $q(V)$, $\sigma_q(V)$) is said to be an *ur-particle milieu of q over V*. We can say that the observable existence of q over V is the triple ($\delta^q(V)$, $q(V)$, $\sigma_q(V)$) in the ur-particle state. Similarly, the triple (δ^q, q, σ_q) is said to be an *ur-wave milieu of q*.

(3) Consequently, when q is not reified at any object of t-site S, i.e., when q is in an ur-wave state, it is meaningless to discuss the position of q, whose particle Q is in a wave state. For instance, such an expression as "A particle Q is *everywhere* when Q is in a wave state" is meaningless since the associated ur-presheaf q is not reified. When a particle Q is in superposition, t-topos theoretically speaking, any discussion about the location of Q is meaningless. Namely, the temporal topos interpretation implies the following. Not being able to specify a position of Q does not imply Q *is* everywhere. This is because, according to our formulation, the ur-presheaf q is not reified, hence q cannot be observed. That is, an expression such as 'An electron corresponding to ur-presheaf *e* can *be anywhere*' is a linguistically inappropriate phrase to describe the position of non-reified electron *e*. For an electron associated ur-presheaf *e* to 'be' at a place, the generalized-time object of t-site needs to be specified. If ur-presheaf *e* is not measured, e.g., until a later time, the choice of t-site objects has not been made yet. Consequently, ur-presheaf *e* is in an ur-wave state until measured. That is, such an expression as 'an electron *can be anywhere* until it is measured' is a contradictory sentence. When electron ur-presheaf *e* is in an ur-wave state, the position of electron *e* cannot be known since a site object is not specified by the very definition of being in an ur-wave state.

The next axiom [*iii*] is a mathematical formulation for expressing a certain state and temporally precedes the other state. Since all the ur-particle states are determined by the parameter domain category S, we need to define a concept in t-site of temporally linearity in terms of a morphism in t-site S. This notion is relevant to the concept of a light-cone in special relativity theory, whose t-site version will be formulated in 2.3, Temporally Linear Morphisms, in this chapter.

We also call objects of t-site *generalized-time* (*periods*). A t-site S plays, as the category of generalized-time periods, the role of a parameter category for all the ur-particle states of any object in t-topos S^. Even though a terminal object $\omega = (\kappa, \tau)$ of t-topos S^ associated with spacetime is also parameterized by objects of t-site S, ω is a constant ur-functor. Note that κ is the ur-presheaf associated with space, and τ is the ur-presheaf associated with time. Then the commonly used parameter 't' for *time* corresponds to τ (V), $V \in Ob$(S) in the following sense. For an object q in S^, there exists a unique morphism σ_q from q to a terminal object $\omega = (\kappa, \tau)$ of S^ by the definition of a terminal object ω. Then morphism $\pi_\tau \circ \sigma_q: q \to \tau$, evaluated at V of S, is corresponding to the usual 't' *associated with presheaf* q, where $\pi_\tau$is the projection from $\omega = (\kappa, \tau)$ to the second coordinate, i.e., $\pi_\tau(\omega) = \tau$, indicating the usual time $t \leftrightarrow \tau = \tau(V)$ is dependent on a particle associated presheaf q. Furthermore, this observation indicates that *time* is a local matter, i.e., depending upon each q in S^. Also notice that since τ is a constant ur-functor, i.e., we have $\tau = \tau(V)$.

[iii] For an ur-presheaf $q \in Ob(S^\wedge)$, let $V \to U$ be a morphism in a t-site S over which q is reified. Then let ur-particle states $q(V)$ and $q(U)$ be the corresponding two ur-particle states of q over V and U, respectively. Let us assume that the ur-particle state $q(V)$ temporally precedes the ur-particle state $q(U)$. Then morphism $V \to U$ is said to be *temporally linear*, which is abbreviated as *t-linear*. We also say that U is *within the ur-lightcone* (or *ur-light cone*) *with vertex V*.

Consequently, the temporal site S is a background temporal category for the usual time 't'. It is our view that objects in t-site play a more fundamental role as parameters than the usual time. We express the above correspondence by

$$t \leftrightarrow \pi_\tau \circ \sigma_q$$

which can be read as:

Time is a concept associated with an individual (i.e., each q).

See Chap. 5 for a philosophical aspect about time and t-site S. See [Barbour (1999)] and [Butterfield-Isham (1999), (2001)] concerning their approaches toward *time*.

The next axiom [*iv*] is to formulate shortest time intervals in terms of t-site objects and morphisms. In other words, we will give a temporal theoretic definition of the shortest time period.

[iv] A morphism $\varphi: V \to W$ in t-site S is said to be a micromorphism when the following condition is satisfied. For an arbitrary factorization of φ as $\varphi = f \circ g$ in diagram

$$\begin{array}{ccc} V & \to & W \\ \downarrow & \nearrow & \\ U & & \end{array} \tag{2.1}$$

where $g: V \to U$ and $f: U \to W$ of $\varphi: V \to W$, if either g or f is an isomorphism, then $\varphi: V \to W$ is a *micromorphism*.

Notes

(1) Intuitively speaking, a micromorphism is the 'shortest' unit as a morphism in a t-site which is impossible to be decomposed into shorter morphisms.

(2) We define the notion of a *microfactoring* of a morphism as follows. A *microfactoring* of morphism $V \to W$ is a factorization of a morphism into micromorphisms. That is, a microfactoring is a decomposition of $V \to W$ into a sequence of micromorphisms. To be precise, consider a factoring of a morphism $\phi: V \to W$ by n objects $\{V_i\}$ between V and W, $1 \leq i \leq n$, as shown in (2.2)

$$V \to V_1 \to V_2 \to \cdots\cdot \to V_n \to W \tag{2.2}$$

If all the morphisms $V_i \to V_{i+1}, i = 0, 1, 2, \ldots, n$, are micromorphisms, then such a factoring in (2.2) of a morphism φ: $V \to W$ is said to be a *microfactoring* of φ: $V \to W$, where $V_0 = V$ and $V_{n+1} = W$. Then we say morphism φ: $V \to W$ has the *microlength of n.*

The notion of a Planck-factorization or a Planck-factoring of a morphism is defined as follows:

Definition For a factoring $V \to U_1 \to U_2 \to \cdots \to U_m \to W$ of $V \to W$, if each morphism h_i:$U_i \to U_{i+1}, i = 0, 1, 2, \ldots, m$, corresponds to a *Planck morphism*, we say that the above sequence (2.2) is a *Planck-factorization* of a morphism $V \to W$. Note that h_i:$U_i \to U_{i+1}$ is a *Planck morphism* when the difference between $t_{i+1} \leftrightarrow \tau(V_{i+1})$ and $t_i \leftrightarrow \tau(V_i)$ corresponds to Planck time which is approximately 5.39×10^{-44} s. Note that if morphism $V \to W$ is a micromorphism, then it is a Planck morphism. The notion of a micromorphism gives a description of the shortest possible change of ur-states.

Next, we need an axiom for a general process for decomposing a large object into smaller objects. We need to formulate a mathematical expression how to break the large object into the smaller objects. Specifically, *for mathematically inclined readers*, the process of decomposition of a macroscopic object into microscopic objects is expressed by describing the ur-presheaf q in terms of ur-subpresheaves $\{q_i\}$ of q, viz., q is a co-product of $\{q_i\}$, i.e.,

$$q_i \to \coprod q_i \stackrel{\text{def}}{=} q.$$

The above morphism $q_i \to q = \coprod q_i$ is interpreted as a subobject q_i being '*embedded*' in the object q. This is the process of decomposing a big entity q into the small parts $\{q_i\}_i$, where i belongs to a finite index set I, i.e., $i \in I$. Furthermore, we can continue this process of decomposition, namely, expressing each q_i as a coproduct of even finer ur-subpresheaves $\{q_{i,j}\}_j$, that is, into even finer $\{q_{i,j}\}$ than q_i. After a finite step of these processes, we reach the stage where we cannot decompose anymore; such an ultimate entity q_{pl} corresponds to the notion of an elementary particle where subscript *pl* comes from *Planck*. That is, such an entity q_{pl} has no smaller entities than q_{pl} itself. We can define such an ur-subsheaf q_{pl} as follows. If a decomposition of q_{pl} were to be performed as further subobjects $\{q'\}$, then q_{pl} is isomorphic to q'.

The other direction from microcosm to macrocosm is performed through the sheaf property (or stack property) by gluing local data. Such a gluing is accomplished through equalities or isomorphisms among 'overlapping' local data. A mathematical expression for such an overlapping is the categorical notion of a *fiber product*.

Let us be more precise:

[v] Let $q \in Ob(S^\wedge)$ be an arbitrary object of t-topos S^. When q can be written as $q = \coprod q_j$, a coproduct of $\{q_j\}$, where $q_j \in Ob(S^\wedge)$ and j belongs to a finite index set J, the coproduct $q = \coprod q_j$ is said to be a decomposition of q. Let us denote the canonical morphism as:

$$\varepsilon_j : q_j \to \coprod q_j = q. \tag{2.3}$$

Furthermore, let $q_j = \coprod q_{i,k}$ be a decomposition of q_j and let the second canonical morphism be:

$$\varepsilon_{i,k} : q_{i,k} \to q_i = \coprod q_{i,k}, \tag{2.4}$$

where $q_{j,k} \in Ob(S^\wedge)$ where j and k belong to finite index sets J and K, respectively. We can continue the process of decomposing $q_{j,k}$ further. When decomposing processes become stable as $q = \coprod q_{j,k,\ldots,l}$, such a decomposition is said to be a *microdecomposition* of q. Note that a decomposition is said to be *stable* when all the ur-presheaves are isomorphic to the next level of decomposed ur-subpresheaves. We call such ultimate ur-subpresheaves *elementary ur-presheaves*.

Note We can express the notion of a microdecomposition of q as the *inverse limit* of $\{\coprod q_{j,k,\ldots}\}$ as the collection of elementary ur-presheaves for $q \in Ob(S^\wedge)$.

Note that q represents an initial large system, namely a global object, and an individual q_j represents a smaller subobject of $q = \coprod q_j$.

When an individual q_j is observed by $p \in Ob(S^\wedge)$, there exists an observation morphism (a measurement morphism) s_j from q_j to p over V and W, respectively, i.e.,

$$s_i : q_i(V) \to p(W). \tag{2.5}$$

Then notice that it is not possible to compose s_j with ε_j, namely, there does not exist a morphism from the large system $q = \coprod q_j$ to observer p. The impossibility of composition of morphisms ε_j: $q_j \to \coprod q_j = q$ and s_j: $q_j \to p$ is clear from diagram (2.7).

On the other hand, when p observes the global system $q = \coprod q_i$, via an observation morphism

$$s : q = \coprod q_j \to p, \tag{2.6}$$

then we can compose s: $q = \coprod q_j \to p$ with ε_j: $q_j \to \coprod q_j = q$ as

$$s \circ \varepsilon_j : q_i \to p,$$

which can be observed from the following commutative diagram:

$$\begin{array}{ccc} q = \coprod q_j & & \\ \uparrow \varepsilon_j & \searrow s & \\ q_j & \to & p. \end{array} \tag{2.7}$$

Then the assignment for morphism s: $q = \coprod q_j \to p$ to morphism $s \circ \varepsilon_j$: $q_j \to p$ can be considered as the restriction of the global system $q = \coprod q_i$ to each smaller q_j, matching the local given data $\{q_j\}$. This is the process from the big to the small, i.e., descending from the global to the local data.

The impossibility of composing morphism s_j with morphism ε_j explains why seeing a small local entity does not tell what the global entity is. The composability of morphisms shows the the interplay between the macrocosm observation and microcosm observation.

2.2 Microcoverings

Our approach toward the concept of time 't' is in terms of objects of temporal site S. This means that we consider even time 't' is parameterized by the objects of temporal site S. We express the correspondence among them as:

't' $\leftrightarrow$ $\tau(V)$ *where* τ *is the temporal ur-presheaf* $\leftrightarrow$ *object* V *of* S.

As a more fundamental concept of a '*true time*' period, we now focus on temporal site S.

We look for a phrase for a shorter time period in terms of coverings of an object in a t-site S. For the definition of a covering, see Appendix II.

Let

$$\{V \leftarrow V_i\}, i \in I \tag{2.8}$$

be a covering of V in S, where ε_i:$V_i \to V$ is a covering morphism.

Furthermore, consider a covering of V_i by another finer covering $\{V_{i,k}\}$, where $\varepsilon_{i,k}$:$V_{i,k} \to V_i$ as

$$\left\{V_i \leftarrow V_{j,k}\right\}, i \in I, k \in K. \tag{2.9}$$

After a finite step of the repeated procedures by finer coverings, the process becomes stable, i.e., all covering morphisms ε's become isomorphisms. Then such a covering is said to be a *microcovering* or a *Planck covering* of V, and each object $V_{i,k,\ldots,l}$ of such a microcovering is said to be a *generalized micro-time period* or an *ur-Planck object* of t-site S. We let $V_{pl} = V_{i,k,\ldots,l}$.

Notes

(1) Classically, a covering of an open set V is a collection of open subsets $\{V_i\}$ of V satisfying $V = \cup V_i$.

(2) *For mathematically inclined readers*, an ur-Planck object V_{pl} can be regarded as the *inverse limit* of $V \leftarrow V_{i,k,\ldots}$ for $V \in Ob$(S).

2.3 Temporally Linear Morphisms

We revisit axiom (*iii*) on linearity for a morphism in t-site S. Let us express the general formulation for communication between two entities represented by presheaves q and p. Note that such entities need not be living creatures. For arbitrary objects p and q in t-topos S^, i.e., $p, q \in Ob(\mathrm{S})$, let $\varphi: V \to U$ be a morphism in t-site S. Suppose that p and q are reified at $V, U \in Ob(\mathrm{S})$, respectively. If there is a noncanonical morphism,

$$s^V{}_U : p(V) \to q(U), \tag{2.10}$$

then we call the non-canonical morphism $s^V{}_U$ an *observation or measurement morphism* of p over V by q over U. Then morphism φ: $V \to U$ is said to be a *temporally linear* (abbreviated as *t-linear*) morphism. Note that p is the observed, and q is the observer. If such an observation morphism $s^V{}_U{:}p(V) \to q(U)$ exists, then we also say that two ur-particle states $p(V)$ and $q(U)$ are *mutually in the same ur-lightcone*. The notion of an ur-light cone is analogous to the traditional notion of a light-cone in special relativity. This is because particle ur-state $p(V)$ is observable by particle ur-state $q(U)$. Also, such a t-linear morphism φ: $V \to U$ in the t-site is said to be in the same *t*-light cone. Then t-site objects U and V are also said to be *mutually in the same ur-lightcone*. Note that this definition of an ur-lightcone differs from the definition given in terms of the time associated presheaf τ. See, for example, [Kato (2013)]. This current definition of an ur-light cone seems to be more universal and elegant than the earlier one.

Notice also that the only difference between the observed and the observer is the direction of a morphism.

As the two sides of fundamental diagram (*), note that we always have the commutative diagrams:

$$\begin{array}{ccc} & \omega & \\ \nearrow & & \nwarrow \\ p(V) & \to & q(U) \end{array} \tag{2.11}$$

In the physical aspect of our fundamental diagram (*). For the cognitive aspect, we have the following commutative diagram:

$$\begin{array}{ccc} & \alpha & \\ \swarrow & & \searrow \\ p(V) & \to & q(U) \end{array} \tag{2.12}$$

Indicating the observed and the observer are mutually related to a terminal object ω and to an initial object α of S^.

The *ur-light* (or *t-light*) *cone with vertex at V* can be expressed in the following diagram (2.13):

$$\begin{array}{ccccccc} \cdot & \cdots & & \cdot & \cdots & & \cdot \\ \cdot & \cdots & & \cdot & \cdots & & \cdot \\ \nwarrow. & \uparrow \ldots \nwarrow & & \uparrow & \ldots & \uparrow & \nearrow \\ & V_i & \ldots & W & \ldots & V_j & \text{Level - 1} \\ & & \nwarrow \ldots & \uparrow & \ldots \nearrow & & \\ & & & V & & & \text{Level - 0} \end{array} \quad (2.13)$$

where all the morphisms appearing in diagram (2.13) with vertex V are temporally linear with vertex V.

The notion of *multiverse* can be explained in terms of the above ur-light cone in S as follows. Let V correspond to the present time. Suppose $V \to X$ is a temporally non-linear morphism which indicates V and X are not mutually in the same ur-light cone. Then there does *not* exist an observation morphism $p(V) \to q(X)$ for any p and q in S^. The number of temporally non-linear morphisms from V corresponds to the number of multiverses with vertex V.

Notes

(1) For a morphism which is not a *t*-linear morphism $V \to U$, such an observation morphism $s^V{}_U$: $p(V) \to q(U)$ does not exist. This is because the two ur-particle states $p(V)$ and $q(U)$ do not belong to the same *ur-lightcone*. In other words, two states $p(V)$ and $q(U)$ are mutually unobservable.

(2) Whenever there exists an observation morphism $s^V{}_U$: $p(V) \to q(U)$, special relativity implies morphism $V \to U$ must be a temporally linear morphism, viz., the t-linearity in a t-site is a necessary condition for an observation morphism to exist.

Remark Our choice of terminology of a *temporally linear morphism in a t-site* is a t-topos theoretic interpretation of the linear nature of the usual time concept and is motivated by the notion of a light-cone.

2.4 Mnemonic Morphism as a Non-canonical Morphism

For any morphism

$$f : V \to U \, in \, t - site \, S,$$

a *contravariant* functor p canonically induces the morphism shown in (2.14), by *reversing* the direction of the arrow from V to U

$$p(f) : p(U) \to p(V). \quad (2.14)$$

See the brief description of a category in the Introduction or Appendix I.

We are ready to introduce a non-canonical morphism, viz. without reversing the direction of an arrow. If p is observed or measured by q over the same generalized-time period V which corresponds to the present, we have a non-canonical observation (measurement) morphism g from $p(V)$ to $q(V)$ so that the following diagram is commutative:

$$\begin{array}{ccc} p(U) & \rightarrow & p(V) \\ & \searrow & \downarrow \\ & & q(V), \end{array} \tag{2.15}$$

where the commutativity of diagram (2.15) means that the measurement of the present state $p(V)$ by $q(V)$ via g receives some information of the future state $p(U)$ via the composite morphism $g \circ p(f)$ from $p(U)$ to $q(V)$. For $p(f)$, see (2.14).

Note However, as indicated in the above (2) in Notes in 2.3, for the special relativity view, $V \to U$ must always be temporally linear.

In particular, for a conscious entity q, such a non-canonical observation morphism (compare with (2.14))

$$s^{V}{}_{U} : q(V) \to q(U)$$

is said to be a *mnemonic morphism* which will play important roles in Chaps. 4 and 5. See also [Kato-Nishimura (2021)].

As an example, let us consider recognition as a psychological phenomenon. Consider person Q meeting another person P for the first time. During the first several minutes of communication between P and Q, Q may get only a first impression of P. After several meetings with P and communicating with P, Q may have a better understanding of P. Then, after e.g., a few years, Q may have a more solid understanding of P in the sense that previous partial understandings of P do not contradict with such a solid understanding after a few years of communication. A question is how we can phrase and formulate such a change of processes of Q understanding P better and better.

For mathematically inclined readers, we can express this flow of information from P to Q in terms of ur-presheaves as follows without considering special relativity in the sense that we use the same object V_0 of t-site in (2.10), (2.11), and (2.12). Namely, the same object V of t-site in (2.15) is used for the ur-particle states for both ur-presheaves p and q, indicating a simultaneous communication between p and q. Let p and q be the corresponding ur-presheaves to P and Q. Then we write the initial contact of Q with P, satisfying the commutativity $\delta^{q}(V_0) = s_0 \circ \delta^{p}(V_0)$ in the following diagram, where $s_0 = s^{V}{}_{V}: p(V_0) \to q(V_0)$ is an observation morphism of p by q over V_0:

$$\begin{array}{ccc} & \alpha & \\ \swarrow & & \searrow \\ p(V_0) & \rightarrow & q(V_0) \end{array} \tag{2.16}$$

where V_0 is an object of t-site S, i.e., letting the common $V = V_0 \in Ob(\mathrm{S})$. Several-minute intervals following the first meeting may still not provide Q a clear understanding of the personality of P. Let us express the following several-minute intervals in terms of objects of temporal site S as $V_1, V_2, \ldots, V_n$. The communication between P and Q can be expressed as the following commutative diagram, i.e., $\delta^{q}(V_i) = s_i \circ \delta^{p}(V_i)$, where s_i: $p(V_i) \to q(V_i)$, in terms of the corresponding ur-presheaves p and q as.

$$\begin{array}{ccc} & \alpha & \\ \swarrow & & \searrow \\ p(V_i) & \to & q(V_i) \end{array} \tag{2.17}$$

where $V_i \in Ob(\mathrm{S})$ for $i = 0, 1, 2, \ldots, n$. That is, P was observed n intervals by Q. The tilted morphisms from α to $p(V_i)$ and $q(V_i)$ are the uniquely determined morphisms $\delta^{p}(V_i)$ and $\delta^{q}(V_i)$, respectively.

However, after several hours, Q may get reasonably accurate information about P. Then the information data consist of those several-minute intervals being considered as local trivial data. The accurate information on P after many hours, or many years, is a non-discrete global understanding of P by Q. Such a global understanding means that further information flow from P to Q becomes *stable. The observation* (or *measurement*) *stability* at $i = n$ is defined as follows. For any further observation of P by Q beyond $i = n$, e.g., $i = m$, $m > n$, all the 'restriction' of $\delta^q(V_m)$ to each earlier state $\delta^q(V_i)$, $n \leq i < m$, is the same as at the level $i = n$. We need to explain 'restriction' in the above sentence. First see the commutative diagram.

$$\begin{array}{ccc} & \alpha & \\ \swarrow & & \searrow \\ q(V_i) & \to & q(V_m), \end{array} \tag{2.18}$$

where k_q: $q(V_i) \to q(V_m)$ in (2.18) is a non-canonical morphism called a *mnemonic morphism*. Then each earlier state $\delta^q(V_i)$ is the restriction of $\delta^q(V_m)$ when $\delta^q(V_m) = k_q \circ \delta^q(V_i)$ holds in diagram (2.18), i.e., diagram (2.18) is commutative.

Our attempt is to unify cognitive activity of a conscious entity with physical constraints coming from quantum and relativistic physics. As noted earlier, the uniquely determined morphisms δ^q: $\alpha \to q$ and σ_q: $q \to \omega$ are essential. Namely, our focus is not on the object q itself but on the morphisms δ^q and σ_q. The natural transformation (i.e., the morphism of functors) $\delta^q(V)$ over a temporal site object V, is the cognitive ur-particle state of q over $V \in \mathrm{Obj(S)}$. On the other hand, σ_q should be understood as the physical, e.g., gravity, influence $\sigma_q(V)$ of q on a spacetime ur-presheaf ω over $V \in Ob(\mathrm{S})$.

Fundamentally speaking, all the physical aspects of t-topos methods involving a spacetime terminal object ω of t-topos S^, i.e., the right-half of diagram (*), come from the consequences of functorial (canonical) properties as shown in [Kato (2004, 2005, 2006, 2010, 2011, 2013, 2017)]; however, an action of measurement is non-canonical, i.e., non-functorially α-induced. When we deal with cognitive aspects of t-topos involving an initial object α, , i.e., the left-half of (*), we need a notion of a non-canonical (non-functorial) morphism, called a *mnemonic* morphism as a self-observation. For a temporally linear morphism $V \to W$ of S, where $V, W \in Ob(\text{S})$ and an object $q \in Ob(\text{S}^\wedge)$, we have the canonically induced morphism $q(V) \leftarrow q(W)$ by the contravariant functor q. However, a mnemonic morphism is given by

$$k_q{}^V{}_W = k^V{}_W : q(V) \to q(W) \tag{2.19}$$

See [Kato-Nishimura (2020, 2021)]. Recall that since an initial object of a topos is a constant functor, functor α is independent of a t-site object. Hence, we have α $\underline{\underline{\text{def}}}$ α(-) = α(V) for an arbitrary object V in S.

Note The notion of a *mnemonic* morphism is a special case of a *measurement* (*observation*) morphism $m(V) \to q(W)$ for the case where two objects in S^ are the same, i.e., $m = q$. See [Kato, 2006, 2008, 2013] for a measurement morphism related to quantum physics, e.g., a double slit experiment.

Definition When the diagram.

$$\begin{array}{ccc} & \alpha & \\ \swarrow & & \searrow \\ q(V) & \to & q(W) \end{array} \tag{2.20}$$

is commutative, the pair (δ(V), δ(W)) is said to be *relevant with respect to* the mnemonic morphism $k^V{}_W$: $q(V) \to q(W)$. Then the mnemonic morphism $k^V{}_W$: $q(V) \to q(W)$ is said to be a *relevant mnemonic morphism for the pair* (δ(V), δ(W)). We will return to this topic in Chap. 4.

Notes

(1) When the above diagram is not commutative, one could interpret non-commutativity as two thoughts δ(V) and δ(W) over V and W, respectively, unrelated, i.e., irrelevant.

(2) The existence of a *relevant mnemonic morphism* is a *t-topos theoretic generalized characterization of an artificial intelligence machine* in the sense that q has an ability to recall (or remember) the past $q(V)$ for the present ur-state $q(W)$. Note also that the choice of the word 'mnemonic' comes from the word 'recalling,' which is expressed as the morphism $k^V{}_W$. That is, mnemonic morphism $k^V{}_W$ connects the past cognitive state δ(V) of consciousness of q to the present consciousness δ(W).

Remark *For mathematically inclined readers*, as a general remark, presheaf F is determined by the objects $\{F(V)\}$ as the 'values,' i.e., objects in codomain category, for all objects V of a t-site S. Then *the Yoneda Lemma* implies that the knowledge of $F(V)$ is equal to.

$$\text{Hom}_{\text{S}^\wedge}(V^\wedge, F),$$

where $V^\wedge = Hom_S(-, V) \in Ob(S^\wedge)$. Furthermore, a convenient notation of $F(G) = F^\wedge(G)$ for $Hom_{S^\wedge}(G, F)$, $G \in Ob(S^\wedge)$ indicates the knowledge of F at G. See, e.g., *Notices of the AMS*, Vol. 64, No. 4, 2016, p. 405. For references for Yeneda Lemma, see [Gelfand-Manin (1996)], [Kato (2006)] or Appendix II.

The restriction from a global object (global understanding in this example) to each local datum (in this case earlier partial understanding) is the descending process from the large to the small.

Here is another example from music: person Q is only allowed to listen to two notes at a time of a well-known music piece B. The procedure is the same as the above example: replace the observed P with the observed B. In this example, we need to formulate 'What understanding is' in terms of commutativity of diagrams, and 'What recognition means' in terms of the restrictions of the above example.

Next consider a (well-known) large painting G. Person Q is allowed to see only a small portion of the painting G at a time, but after a while, if those small portions cover a large enough area or the entire painting G, then Q will know what painting G is, assuming Q's ability to obtain a global object by 'gluing' all those locally shown portions.

Readers are recommended to provide explicit formulations for these preceding three examples.

In physics, macroscopic physical entities consist of atomic particles. Our experiences since childhood consist of observing and feeling the space–time relationship among macroscopic objects. For example, we have learned how to catch and hit a ball as part of the game of baseball. Galileo and Newton encoded the principles of bodies in macrocosm affected by gravity. Einstein found much deeper meaning of space–time relationships affected by gravity.

The phenomenon relating the probabilistic microscopic world of electrons, quarks, leptons, bosons, … to the deterministic macroscopic world of living creatures, stones, planets, stars, …is called *decoherence* in physics. In mathematics the *sheaf* (or *descent*) *theory* explains the relationship between local data (microcosm) and a global object (macrocosm). Since microcosm is a part of macrocosm, i.e., they are not separated from each other, we should not be happy with an approach to describe *one universe* with two different theories. Namely, one theory is restricted to microcosm, and another theory is restricted to macrocosm. For example, an emerged living entity from local non-living parts should be obtained by organizing local data. Furthermore, the restriction of the global entity to each local datum should coincide with the locally given entity.

In life sciences, situations are much more complex since locally trivial data can consist of atoms and molecules which ultimately consist of large numbers of elementary particles, but the induced self-organized global object from those non-living pieces may be a living creature. We must answer such a question as:

What is an appropriate definition for the characterization for a living entity?

and

What is an animate entity which is distinguishable from an inanimate entity?

by providing an explicit mathematical formulation. To answer the above questions, our approach requires an initial object α of t-topos S^. We are assuming that an individual elementary particle, e.g., an electron, in a stone is identical (isomorphic as ur-functors) to an electron in a human brain. First, these electrons are indistinguishable in the following sense.

All the ur-presheaves associated individual electrons are isomorphic but are not equal

Second, for the case of a human being, the pyramid of organization occurring in our body from the small trivial entities to the large entity can be expressed as progressing from the simplest to more complex: elementary particles like electrons and quarks $\mapsto$ atoms $\mapsto$ molecules $\mapsto$ cells $\mapsto$ human. One can extend this process as human $\mapsto$ society $\mapsto$ ..., etc. The advancement is not only in scale increase from smaller to larger, but also it is a process for organizing from structurally fundamental and simple entities to more complex emerged objects. As we noted earlier, the process from the trivial objects to a non-trivial object is our fundamental theme. We do this process by gluing (via either equalities or isomorphisms) over the 'overlaps' of the trivial data to obtain a non-trivial global object.

2.5 From Mathematics

As a common theme of those above examples, we need to provide formulations for the process from the small to the large by 'gluing' local data to obtain a global object. This is the very matter of the modern mathematical field in algebraic geometry and complex analytic geometry, i.e., the sheaf theoretic methods, which was mostly developed around WWII and further developed in 1950's.

Let us review what we have described in Chap. 1 and up to this point.

A key question becomes:

Can we glue local data on the 'overlaps' to obtain a global object?

Historically, we have been asking the following questions, as examples:

What is a rock made of, and what does air consist of?

What are the most fundamental elements of all emerged macroscopic entities?

Various methods have been developed to answer those questions in physics, chemistry, and biology.

One of the most basic approaches among scientific analyses is to decompose a macroscopic object into more structurally simpler and fundamental smaller entities. Namely, by decomposing a large object m into smaller objects $m_1, m_2, \ldots$, then we say that m consists of $\{m_1, m_2, \ldots\}$. Then we study these smaller and more fundamental $\{m_i\}_i = \{m_1, m_2, \ldots\}$ by asking what those decomposed pieces are. The main theme of this treatise is the reverse question. Namely, the direction of our study is to build a global macroscopic object m from a collection $\{m_i\}$ of more fundamental and smaller objects. Then we need to confirm that the restriction of such obtained m, which is obtained by gluing local data, to each local object coincide (i.e., being isomorphic to) with the original m_i. Then we need to answer the following questions.

(1) *How do we glue local data* $\{m_i\}$*?*
(2) *What is a mathematical definition for the restriction of the large object to each datum?*
(3) *Furthermore, how do we capture the stage change from inanimate objects to such a glued animate object?*

We consider (3) to be crucially important.

We also include nonmaterial objects. For example, as locally trivial entities $\{m_i\}$, we can take individual musical notes. A composed music piece m consists of those individual musical notes $\{m_i\}$, where such an individual note m_i is a trivial musical note, i.e., only a sound. On the other hand, non-trivial global object m is an emerged piece of music from local data of individual musical notes. The reverse direction of obtaining a non-trivial object m from trivial local objects $\{m_i\}$ is more delicate. The notion of a sheaf is used to formulate the process of gluing local data.

2.6 Gluing Through Isomorphisms

Let q be an object of S^ and V be an object of S. Then let $q = \coprod q_j$ be a decomposition of q and let $\{V \leftarrow V_i\}_i$ be a covering of $V \in Ob(\mathrm{S})$. Suppose that q_j is reified over V_i. Let us assume that ur-presheaf q_j is reified over V_j after rearranging the indices.

Consider a pair (q_j, φ_{ij}) where $q_j(V_j)$ and $q_i(V_i)$ are reified. Over $V_i \times_V V_j$, morphism φ_{ij} is an isomorphism from q_i to q_j, i.e., from $\pi_i^* q_i$ to $\pi_j^* q_j$.

We call φ_{ij} a *gluing isomorphism* between the ur-presheaves q_j and q_j over $V_i \times_V V_j$, i.e., an isomorphism between $\pi_j^* q_j$ and $\pi_i^* q_i$, π_j^* over $V_i \times_V V_j$. Note that π_j^* is the inverse image (the 'restriction' in the case of a topological space) of π_j. See [Gelfant-Manin (1996)] or [Kashiwara-Schapira (2006)] for the notion of inverse image. A pair $(\{q_j\},\{\varphi_{ij}\})$ is called descent data.

Note that $V_i \times_V V_j$ is a fiber product of V_i and V_j over V, i.e., the following diagram is commutative:

$$\begin{array}{ccc} V_{ij} \overset{\text{def}}{=} V_i \times_V V_j & \xrightarrow{\pi_j} & V_j \\ \downarrow \pi_i & & \downarrow \\ V_i & \rightarrow & V, \end{array}$$

where π_i and π_j are the projections from the fiber product $V_i \times_V V_j$ to V_i and V_j, respectively. Then the restriction of a global object q reified over V is obtained by the inverse image $h_i^* q$ of q under $h_i: V_i \rightarrow V$. When restriction $h_i^* q$ is isomorphic to each q_i, an ur-presheaf q is said to be a global object obtained from local data $\{q_i, V_i\}$.

For mathematically inclined readers, diagram.

$$
\begin{array}{cccccc}
 & V_{ijk} & \rightarrow & & V_{jk} & \\
 & \swarrow\downarrow & & & \swarrow\downarrow & \\
V_{ij} & & \rightarrow & V_j & & \\
\downarrow & V_{ik} & \rightarrow & \downarrow & V_k & \\
 & \swarrow & & & \swarrow & \\
V_i & & \rightarrow & V, & &
\end{array}
$$

where $V_{ijk} = V_i \times_V V_j \times_V V_k$, may be useful. The gluing isomorphisms satisfy the 'restrictions' of $\{\varphi_{ij}\}$ to $V_{ijk} = V_i \times_V V_j \times_V V_k$. Namely, $\pi_{ij}{}^{*}\varphi_{ij}$ composed with $\pi_{jk}{}^{*}\varphi_{jk}$ equals $\pi_{ik}{}^{*}\varphi_{ik}$, i.e., meeting the cocycle condition.

2.7 Synopsis

After a short review of Chapter I and the history of the mathematical concept of space, we begin with the axioms for a general topos to be a temporal topos, abbreviated as t-topos. We are providing technical foundations for our treatise in this chapter. Furthermore, the ultimately small objects and morphisms are introduced as ur-Planckian objects which are obtained by decomposing an object in temporal topos S^ and an object in a temporal site S. An ur-Planck morphism as an ultimate morphism in t-site S, which is called a micromorphism, is obtained by microfactoring a morphism in S. Then the notion of a temporally linear morphism in S is introduced which plays a significant role for the concept of an ur-lightcone with vertex V in t-site S. The measurement phenomenon is to be explained in terms of an ur-lightcone with vertex V as present time. The definition of a mnemonic morphism is given as a non-canonical morphism, where a canonical morphism is induced functorially by a contravariant ur-functor. The notion of a mnemonic morphism is applied to cognitive sciences in Chap. 4.

Chapter 3
Temporal Topos and Physics

Contents

It must be admitted that the meaning of quantum physics, in spite of all its achievements, is not yet clarified as thoroughly as, for instance, the ideas underlying relativity theory. The relationship of reality and observation is the central problem. We seem to need a deeper epistemological analysis of what constitutes an experiment, a measurement and what sort of language is used to communicate its result.

Herman Weyl in *Philosophy of Mathematics and Natural Science*, Princeton University Press, 1949, page 264.

The significant problems we face cannot be solved at the same level of thinking that was used when we created them.

Albert Einstein in *The Ultimate Quotable Einstein*, collected and edited by Alice Calaprice with a forward by Freeman Dyson, Princeton University Press, 2011, page 476.

3.1 Background

As quoted in Chap. 1, Grothendieck says (in *Reaping and Sowing*) that the gap from classical physics to quantum physics is like the gap between French and Chinese. We shall observe the gap in this chapter.

G. C. Kato, *Temporal Topos Methods for the Philosophy of Natural Sciences*,
https://doi.org/10.1007/978-981-96-2420-1_3

At the beginning of the twentieth century, Niels Bohr in Copenhagen, Denmark, was a main figure of quantum mechanics during its early development. Let us explain the so-called Copenhagen interpretation by beginning with an unobserved electron, which is said to be described as a state of superposition, meaning the probability of finding the electron is spread out 'over here' and 'over there;' however, according to our t-topos theoretic formulation, it will never be 'everywhere.' The Copenhagen interpretation says that the location of the electron can be specified as 'here' only when the electron is observed. For Bohr, a meaningful proposition about the position of an electron was possible only when it could be expressed using the terminology of classical theory.

John von Neumann gave a mathematical foundation for quantum mechanics based on a Hilbert space. It was also von Neumann who started game theory, was involved with the Manhattan project, and provided the fundamentals for computers. At the Institute for Advanced Study in Princeton, there is a building named ECP for von Neumann. However, due to Einstein's objection to non-pure theoretic research, the ECP building is on the other side of South Olden Lane, i.e., outside the IAS campus. The usual computer of the von Neumann type is based on 2-adic, i.e., 0's or 1's. This approach is very discrete, namely a Yes or No type. On the other hand, as for quantum computing, rather than either Yes or No, a switch can be a sequence of superpositions as a parallel processing.

If asked to construct *one* theory called quantum gravity, from which general relativity and quantum theory both can be derived, one must specify the role played by fundamental units like Newton's constant G, Planck's constant $\hbar$, and the speed of light c.

We note that the Planck length $Pl = 10^{-33}$ cm $= (G \cdot \hbar / c^{3})^{1/2}$, which should be compared with the diameter 10^{-8} cm of an atom, the Planck time $P - t = 10^{-42}$ s $= P\text{-}l/c$, the Planck mass $P\text{-}m$ is 10^{-5} g $= \hbar/c \bullet P\text{-}l$, and the Planck energy $P\text{-}e = 10^{18}$ GeV $= P\text{-}m \bullet c^2$.

Furthermore, we need to address phenomena like wave-particle duality (double slit experiment), the Einstein–Podolsky–Rosen experiment (non-locality), quantum entanglement, the two types of uncertainty (uncertainty principle and uncertainty relation), and the measurement phenomena.

We will briefly comment on the two types of *uncertainty* that are often confused. Indeed, even Heisenberg himself confused them. The so-called *uncertainty principle*, formulated by Heisenberg's microscope thought experiment, is an epistemological inequality that illustrates the trade-off between measurement error and disturbance. On the other hand, *the uncertainty relation* is an inequality derived when two physical quantities are non-commutative. This is unrelated to measurement and is an ontological inequality about the *uncertainty* inherent in the physical quantities themselves. Heisenberg's *uncertainty principle* is a fundamental principle in quantum theory stating that it is impossible to measure precisely both the state of a particle and its rate of change. We explain in more detail later in this chapter about the two types of *uncertainty* in quantum theory.

The analogous notion, in terms of our language, i.e., categories and sheaves, of a light cone must be described together with the dynamic aspect of a gravitational effect both in macrocosm *and* microcosm. Quantum physics was developed to explain the nature and behavior of microscopic entities like atoms and even subatomic particles. However, the theory developed to explain the nature of microcosm turned out to be against our intuitive view from classical physics. For example, even if data are known and provided about

(1) the present location of a particle,

and

(2) how fast a particle is moving, i.e., the particle momentum, the particle's future location and momentum may not be predicted. Furthermore, we must address measurement phenomena. Namely, the quantum mechanics in microcosm is very different from Newtonian mechanics. We must not expect the well-established behavior description and expectation gained and learned from the view of macrocosm when we talk about microscopic entities like electrons, photons, protons, etc. which no longer act like well-behaving localized entities moving on the expected trajectories. It is an incorrect image to consider an electron moving around the nucleus of an atom. The wavefunction, a mathematical tool used in quantum physics, is a very appropriate device for microcosm to describe the behavior of a quantum entity. Even though the wavefunction is still a superb computational tool for the fuzzy quantum reality, it can never be an entity, like a table, manifested at an objective level. However, the ur-presheaf representing an electron may be closer to the quantum reality than a pure computational device, the wavefunction.

Einstein made a comment that

> Quantum mechanics is certainly imposing. But an inner voice tells me that this is not yet the real thing. The theory yields much, but it hardly brings us closer the Old One's secrets. I, any case, am convinced that He does not play dice.
>
> To Max Born, December 4, 1926. In Born-Einstein Letters, 88. Einstein Archives 8–180. The popular version of the last sentence is "God does not play dice with the universe." in The Ultimate Quotable Einstein, collected and edited by Alice Calaprice with a forward by Freeman Dyson, Princeton University Press, page 380.

God does not roll the dice meaning that he objected to the probability aspect of the quantum theory of physics. One reality is chosen among the probabilistic distribution of possibilities of being here and there. As a measurement phenomenon, we ask why the particular choice happens, and what causes the choice.

Since there exists the *uncertainty* for (1) and (2) above for a subatomic entity like an electron, there is no chance to be able to simultaneously predict with certainty the future position and momentum of an electron. We will explain, using an ontological approach called temporal topos, why a particular choice was made among other possible choices in the epistemological issue of measurement phenomena.

Some unexpected behaviors can be observed among subatomic particles. On the other hand, classical physicists like Newton and Einstein encoded the behavior of larger bodies in a macrocosm. We are aware that a macroscopic animate object requires many microscopic objects (e.g., cells, molecule, atoms, ...) to form such an emerged larger animate body, e.g., a cat, in an organized way. Even such a higher functioning entity as self-awareness or self-consciousness is an organized large object composed of the individually trivial small objects. We need to encode explicitly the process from the local trivial objects to a global object, like a brain, as an organized collection of microscopic objects. Namely, our task is to specify how to glue local data to form an emerged global object which is our interpretation of the notion of decoherence in quantum physics connecting a gap of the microcosm and the macrocosm. Namely, the concept of decoherence is to connect quantum phenomena with emerged macroscopic classical phenomena. That is, decoherence is to explain how individual subatomic behaviors can emerge as macroscopic phenomena.

In order to formulate the phenomenon of decoherence in terms of t-topos, we need two ingredients as follows. First, let $q(V)$ be in an ur-particle state of q over V. As the first ingredient, we need a mathematical expression for breaking a large entity q into smaller entities $\{q_j\}$. More explicitly, let q be an ur-presheaf representing an object in macrocosm and let a coproduct of $\{q_j\}$

$$\coprod_j q_j \leftarrow q_i \tag{3.0}$$

be a decomposition into sub-objects $\{q_j\}$ of q in microcosm. Diagram (3.0) should be read that q_j is a part of the whole $q = \coprod_j q_j$.. That is, diagram (3.0) can be interpreted as q consists of a collection of finitely many microscopic parts $\{q_j\}$. As the second ingredient, we need the notion of a covering $\{V \longleftarrow V_i\}_i$ of V. Then decoherence as connecting the quantum world with the macro world is to explain how local data $\{q_j(V_i)\}_{j,\, i}$ relates to an emerged global object $q(V)$. See Appendix B for the notion of a covering.

In this chapter we will provide the t-topos theoretic structure theory, i.e., in terms of categories and sheaves, both for quantum theory and relativity theory.

We provide the t-topos theoretic interpretations and the explicit formulations for such quantum physical phenomena as:

The wave-particle duality and superposition principle

The non-locality and quantum entanglement

The concept of uncertainty

The measurement problem *and multiverse interpretation*

The issue of determinism and free will.

We will first review and rephase how we have interpreted various phenomena mentioned above in terms of our temporal topos theoretic methods. Also see [Isham (2010)].

When we want to specify the location and the time of an object on earth, we traditionally use longitude, latitude, altitude, and time for 'where and when.' Because of this specification, we often assign the coordinates (x, y, z, t) for 'where and when.' Our approach is to assign a ur-presheaf $\omega = (\kappa, \tau)$ for space and time rather than the coordinates (x, y, z, t). In our ur-presheaf expressions, we have the correspondence:

$$(x, y, z) \leftrightarrow \kappa$$

$$t \leftrightarrow \tau.$$

To indicate the position (x, y, z) of a particle P, we need to specify 'when', i.e., depending upon time. We express the position as $(x(t), y(t), z(t))$, indicating the coordinates depending upon time t as the parameter.

In general, since an ur-presheaf is an assignment from a category to a category, we need to specify the domain category where those ur-presheaves are defined. Such a domain category S where ur-presheaves are defined, is said to be an ur-site, i.e., it is a category with a *Grothendieck topology*. See Appendix B for the definitions of these notions. Note that in physics, time t acts a parameter as in the above. In our temporal topos approach, an object V of an ur-site S plays the role of a parameter. Namely, for $p \in Ob(S^\wedge)$ an *ur-state* of p over V is expressed as $p(V)$.

We regard an initial object α as the source of perception, and a terminal object ω as the spacetime background. Then our ontological view can be summarized in the following commutative diagram connecting spacetime, matter, and perception. Namely, for an arbitrary object q of S^, which represents either a microcosm object or a macrocosm object, we have the following fundamental commutative diagram (*) satisfying $\zeta \stackrel{\text{def}}{=} \delta^{\omega} = \sigma_\alpha = \sigma_q \circ \delta^q \colon \alpha \to \omega$.

$$\begin{array}{ccc} \alpha & \rightarrow & \omega \\ & \searrow \quad \nearrow & \\ & q & \end{array} \qquad (*)$$

where $\delta^q \colon \alpha \to q$ and $\sigma_q \colon q \to \omega$ in diagram (*) are uniquely determined morphisms. See Chap. 1 for diagram (*) and Appendix 1 for the definitions of an initial object and a terminal object of a category.

The right-half of commutative diagram (*) can be viewed as the physical world, and the left-half of (*) can be viewed as beyond the physical world, i.e., the cognitive world of perception.

3.2 Gravity

We will provide a formulation for obervation in terms of our language. First, recall that since spacetime ur-presheaf ω is a terminal object of a temporal topos S^, there exists a uniquely determined morphism σ_p from any object p of S^ to ω, by the definition of a terminal object. Our method of temporal topos S^ regards temporal site S as a parameter-category in such a way that the particle ur-state $p(V)$ of p is determined by an object $V \in Ob(S)$. Since ω is a constant functor, we can write $\omega(V) = (\kappa(V),\tau(V)) = (\kappa(-),\tau(-)) = (\kappa, \tau) = \omega$. We focus on the morphism $\sigma_p(V)$: $p(V)$

$\to \omega$, which is regarded as the effect on ω caused by p. When p happens to be an electron representing ur-presheaf e, the gravitational effect σ_e on spacetime ω caused by e is a natural transformation from e to ω, i.e., a morphism of functors.

For a non-canonical observation morphism s: $p \to q$, we have the following commutative diagram in S^:

$$\begin{array}{ccc} & \omega & \\ {}_{\sigma_p}\nearrow & \uparrow_{\sigma_q} & \\ p & \to q. & \end{array} \tag{3.1}$$

Furthermore, when p and q are reified for a t-linear morphism $V \to W$, we have the following commutative diagram:

$$\begin{array}{ccc} & \omega & \\ \nearrow & \uparrow_{\sigma_q(W)} & \\ p(V) \to & q(W). & \end{array} \tag{3.2}$$

When q measures over W the gravitational spacetime effect caused only by p over V, we have the following diagram:

$$\begin{array}{ccc} \omega(V^*) = \omega(\text{-}) & & \\ {}_{\sigma_p{}^V{}_{V^*}}\uparrow & \searrow & \\ p(V) \to q(W), & & \end{array} \tag{3.3}$$

where $s^{V^*}{}_W : \omega = \omega(-) = \omega(V^*) \to q(W)$ is a non-canonical observation morphism of spacetime ω over V^* by q over W. Then the gravitational information on the ur-state of p over V measured by q over W via spacetime $\omega(V^*) = \omega(-) = \omega$ is given by the composition of morphisms

$$s^{V^*}{}_W \circ \sigma_p^V{}_{V^*} : p(V) \to q(W) \tag{3.4}$$

where $V \to V^* \to W$ are temporally linear morphisms. Note that since ω is a constant functor, the specification of an object in S is irrelevant for ω. Consequently, the gravitational effect caused by p over V measured by q over W becomes $s(W) \circ \sigma_p(V)$, which we may write as

$$\mathrm{Grv}^{p(V)}{}_{q(W)} : p(V) \to q(W). \tag{3.5}$$

Note We consider that the independence of V^* for the gravitational formula $\mathrm{Grv}^{p(V)}{}_{q(W)} = s^{V^*}{}_W \circ \sigma_p{}^V{}_{V^*}: p(V) \to q(W)$ in (3.4) and (3.5) (because of the constancy of functor ω) is important; for example, the mass of p should be independent of the distance between the location of p determined by V and the spacetime ω measured over V^* by q over W. That is, the measurement by q of the mass of p via the gravitational effect on spacetime ω over V^* caused by p should have nothing to do with

V^*, e.g., the distance between the location of p determined by V and the space and time determined by V^*.

Next, we will generalize the above case of only the observed p to include the observer q's gravitational effect. Namely, we like to give a formulation for both gravitational effects not only by the observed p but also by the observer q. As q measures the gravitational effect on ω by p, q cannot escape from measuring q's own effect on spacetime ω. That is, we must consider both σ_p and σ_q. As noted, we have uniquely determined morphisms $\sigma_p(V)$: $p(V) \to \omega(V^*) = \omega$, and $\sigma_q(W)$: $q(W) \to \omega(V^*) = \omega$. When q observes the spacetime effect caused by p through an observation morphism $\omega \to q$, an observation of p by q over V and W, respectively, we must consider both effects on spacetime ω from the observed p and the observer q through $\sigma_p(V)$ and $\sigma_q(W)$, respectively. Explicitly, the observation morphism $s^V{}_U$: $p(V) \to q(W)$ consists of compositions of morphisms of spacetime effects of $\sigma_p(V)$ and $\sigma_q(W)$ by both p and q, i.e., $s^{V^*}{}_W \circ \sigma_p{}^V{}_{V^*}$ in (3.4), and the corresponding auto-effect by q is given by

$$S^{V^*}{}_W \circ \sigma_q^W{}_{V^*} : q(W) \to q(W) \tag{3.4auto}$$

for the observation morphism $\omega(V^*) \to q(W)$ of spacetime over V^*. Consequently, the measurement of p over V by the observer q over W is given by both effects from p and q, i.e., of $s^{V^*}{}_W \circ \sigma_p{}^V{}_{V^*}$: $p(V) \to q(W)$ and $s^{V^*}{}_W \circ \sigma_q{}^W{}_{V^*}$: $q(W) \to q(W)$.

That is, in our interpretation, the observed and the observer are mutually related either in macrocosm or in microcosm. That is, an ur-presheaf q cannot measure p without measuring the observer q's effect on a terminal object ω, i.e., $\sigma_q(W)$: $q(W) \to \omega = \omega(V^*)$.

Note Recall first that for a morphism f: $V \to U$ in t-site S, there is a canonically (i.e., functorially) induced morphism by a contravariant functor q, i.e., $q(f)$: $q(U) \to q(V)$, reversing the direction of the morphism in $V \to U$ to $q(V) \leftarrow q(U)$. On the other hand, an observation (or measurement) morphism from p to q and also a mnemonic morphism (which appears in Chap. 4) are non-canonical morphisms satisfying the commutativity of the following diagram:

$$\begin{array}{ccc} & \alpha & \\ \swarrow & \downarrow & \\ p & \to & q. \end{array}$$

See [Kato (2004), (2005), (2010], (2017)].

3.3 The Wave-Particle Duality and Ur-superposition

To be jocular: if a reader is asked, "Are you alive?" the reader responds, "Yes, I am alive." If the reader is dead, there will be no answer. Sometimes no answer is a correct answer. There are no appropriate words from the dead who are not in the world of the living.

A principle states that there are two ways (i.e., a wave or a particle) to describe the same quantum entity, e.g., an electron. Such an entity is either in a particle state or in a wave state. Furthermore, one cannot say such an entity is simultaneously both a particle and a wave; an electron acts either as a particle or as a wave. By introducing a temporal site S as a parameter category, abbreviated as t-site, the phenomenon of wave-particle duality is formulated as whether the ur-presheaf e associated with e.g., an electron, is reified at an object of the t-site or not. We also say that an object p of temporal topos S^ is in *ur-superposition* when p is *not* in an ur-particle state, which can be expressed as $p(-)$, or simply as p.

One of the most important experiments, if not the most, in quantum physics is the double slit experiment, which indicates the nature of the wave-particle duality of a quantum entity.

Even though the setting for this experiment is well known, we will briefly describe this important experiment in quantum physics as follows. Electrons $\{e_i\}_i$ are fired from a source G perpendicularly toward a vertical detector screen D. But there is a barrier Sl between the source G and D, with two parallel slits A and B, which is also parallel to screen D. Let us locate A and B equidistance from the source G. Then the purpose of the double slits experiment is to understand the distribution of marks made by the electrons hitting the detector screen D. Electron e_1 leaves the source G toward D through the double slitted barrier Sl. *After* electron e_1 arrives at D, then the next electron e_2 is fired toward the detector screen D. Thus, we have a steady stream of electrons $\{e_i\}_i$ being fired from G toward the detector screen D, i.e., one electron at a time is fired toward D. We should be aware that each electron is localized at the detector screen D as it hits D, which can be observed by a macroscopic observer like us. That is, we know where and when an individual electon hits screen D.

Then there is a region most frequently hit by the electrons on D, which is about equidistance from slits A and B. The spot most frequently hit on D is neither the spot a nor spot b on D, where a is on the line $G - A - a$, and b is on the line $G - B - b$. We provide an explanation based on temporal topos theory for the distribution of dots on screen D.

This famous double-slit experiment is strange to us as a macro sized creature. Here is one unshakable fact: each electron must go through the barrier Sl, otherwise we have no dots at all on screen D. We have two concerns. The first concern is that judging by the pattern of dots on D even one electron at a time fired arrives at screen D with a self-interference pattern distribution of dots on D. The second concern is which slit, A or B, each electron goes through. According to the macrocosm creature's logic, the answer is neither A nor B. One can close, e.g., slit A. One can do the same by closing B. In both cases, the pattern indicates each electron characteristic of a particle. Then we have only one choice left. Namely, an electron goes through both A and B. However, one electron cannot go through two slits at the same time. Our interpretation for the double slit experiment is as follows. As long as there are two slits, i.e., more than one slit, an object in a t-site S cannot be specified when an electron goes through barrier Sl. Hence, then an electron represented by ur-presheaf e is in a wave state. See [Kato (2004–2017)] for the double-slit interferences in terms of temporal topos.

As we saw in the above, this extremely peculiar phenomenon of the double slits experiment is the essential nature of the quantum realm. An intense dark region on the screen is located directly opposite to half-way between slit A and slit B. The distribution of marked spots on screen D made by electrons is a consequence of being in an ur-superposition of an ur-entity e. The pattern on screen D is caused by the interference effects of wavelike behaving electrons. When each electron arrives on the screen, it is like a particle. On the other hand, the distribution of the interference pattern of the electrons on the screen indicates a wavelike nature. That is, the double slit phenomenon is a perfect example of an electron's wave-particle duality. "Which slit does each electron pass through before hitting the screen?" is a meaningless question in the quantum realm. It is incorrect to say the individual electron went through both slits since if that were the case, there would be two objects V_1 and V_2 so that $e(V_1)$ and $e(V_2)$ are in ur-particle states passing through slit A and slit B, receptively. Then, there should not be a diffraction pattern on the screen.

We also need to pay attention to language. As mentioned earlier, an expression like "A particle P is going through slit A" indicates that there is a moment of time when P passes through A. This very expression is our common language spoken by a macro-object like a human being. When particle P happens to be a micro-object like an electron, we cannot assume that our language is valid to describe a behavior of an electron in the realm of a microcosm.

We rephrase the sentence "Particle P is in a wave state, i.e., P is in the superposition" as "the ur-presheaf p associated with particle P is in an ur-wave state." That is, ur-presheaf p is not reified at any object in the t-site S, i.e., p is in an ur-wave state. We must also be careful about a linguistic expression like "An electron is *everywhere.*" Two things are inappropriate in the expression *everywhere.* One: even a bounded region in three-dimensional Euclidean space $\mathbf{R}^3$ (with coordinates x, y, and z) is finite: there are infinite points in the region. We claim that word '*everywhere*' needs to be replaced with the expression '*at finitely many locations.*' Two: even an expression of '*at finitely many locations*' still indicates at those places there must exist a finitely many t-site objects $V_1, V_2, \ldots, V_n$ over which the ur-presheaf e associated with an electron is reified, i.e., in ur-particle states as $e(V_i)$, $i = 1, 2, \ldots, n$.

Note The quantum state is expressed as a linear superposition

$$|\psi\rangle = u|A\rangle + v|B\rangle.$$

of the two states for position A and position B, where u and v are complex numbers. The rate of change of ψ with respect to time is expressed in the Schrödinger equation.

See [Kato-Tanaka, (2006)].

3.4 The Non-locality and Quantum Ur-entanglement

Quantum entanglement is a phenomenon that a cause at one location seems to allow *immediate* effects at distance places, even outside the light-cone, which is a part of the *Einstein–Podolsky–Rosen Paradox* (often referred to as the *EPR Paradox*). We will specify the meaning of *immediacy* in our language of temporal topos. Based on quantum theory, let two particles P and P^* interact at a point in their history. Let p and p^* be the associated ur-presheaves with P and P^*, respectively. Even when ur-sheaves p and p^* have not interacted, p and p^* seem to be able to still influence each other. This strange effect is said to be the principle of nonlocality. Einstein considered this effect to be 'spooky' and argued that quantum theory is only a partially successful theory to describe the physical universe. Namely, according to Einstein, quantum mechanics is incomplete and wrong since such a 'spooky' effect is unreasonable. This is one of the rare occasions that Einstein was wrong. That is, quantum mechanics both theoretically and experimentally has turned out to be neither incomplete nor wrong. However, the *Einstein–Podolsky–Rosen* paper has provided significant and challenging questions for the next generations to solve. Consequently, the strange and unreasonable nature of quantum phenomena have been confirmed. The principle of nonlocality is a consequence of quantum entanglement. In fact, there is no physical influence between particles represented by ur-presheaves p and p^*, only a correlation due to quantum entanglement. We give a definition of quantum ur-entanglement of ur-presheaves as follows. A pair (p, p^*) of ur-presheaves p and p^* is said to be *ur-entangled* when the pair acts as one ur-presheaf. More precisely, p and p^* are ur-entangled when the following formula holds:

$$\left(p, p^*\right)(V) = \left(p(V), p^*(V)\right). \tag{3.6}$$

Generally, we say that two (or more) entities p and p^* are ur-entangled if the state of one, either p or p^*, has a relationship (correlation) with the other, independent of the distance between p and p^*. According to this definition of quantum ur-entanglement, nonlocality is the same as quantum entanglement. Hence, *the meaning of the immediacy in t-topos* is that an ur-particle state of the pair (p, p^*) is defined by the *same object* V in t-site. Recall that we view an object (a generalized-time period) of t-site as a more fundamental parameter than the usual time concept.

3.5 The Concept of Uncertainty

Generally, the more accurate information we have, the better prediction we get. This is a problem concerning knowledge, namely, it is an epistemological issue. In classical physics, when we have the correct initial conditions, for example, a cannon ball is launched at a determined place and with a certain speed in the chosen direction, we can determine where the cannon ball will be after a few seconds. This is known as mechanistic determinism. Of course, even outside the context of physics, some people believe in fate as a form of determinism. The German poet Rainer Maria Rilke thought the future stands still, and we are the ones who move around the determined future. In fact, this kind of thinking has existed since ancient times, and it can be said that Aristotle's notion of the 'final cause' also aligns with a similar worldview.

Then we want to ask, "How about free will?" Even those macrocosm events with the involvement of free will, our future seems to be flexible. In the microcosm, things are more complex since we cannot pinpoint the position of a subatomic particle like an electron. *This is the issue of uncertainty in quantum theory, and furthermore, there are two types of uncertainty as conceptual problems.*

We must pay attention to two conceptually distinct types of *uncertainty*: *the uncertainty principle* and *the uncertainty relation*. Strictly speaking, what Heisenberg discovered was *the uncertainty principle* not *the uncertainty relation*. What conceptual differences exist between the two uncertainties?

In quantum theory, *uncertainty* can be broadly categorized into two types:

1. *Epistemological uncertainty*:
   Indicates the limits of measurement in quantum systems.
2. *Ontological uncertainty*:
   Demonstrates the inherent *uncertainty* of physical quantities.

Actually, Heisenberg's *uncertainty principle* regarding measurements was experimentally violated in 2012. Of course, it is important to clarify that it was *the uncertainty principle*, not *the uncertainty relation*, that was violated in the 2012 experiment. While this is widely known within the research domain of quantum foundations, it is not generally well-known. Therefore, to be clear, let us explain it.

The confusion originates with Heisenberg himself. Light consists of tiny particles called "photons." When a large number of photons are directed at an electron and the scattered photons are captured by a microscope, they create a blurred image with a certain degree of dispersion. Consequently, the exact position Q of the electron cannot be determined. The magnitude of this dispersion is proportional to the wavelength of light, which becomes the measurement error $\varepsilon(Q)$ for position. Furthermore, when an electron scatters a photon, the electron itself is also affected by the photon, causing a change in momentum P. This disturbance in momentum is called $\eta(P)$. Although this disturbance occurs randomly each time, on average, it is inversely proportional to the wavelength. In 1927, from his famous thought experiment with a gamma-ray microscope, Heisenberg anticipated that the relationship between the error and disturbance would be approximately

$$\varepsilon(Q)\eta(P) \sim h,$$

which became known as the original *uncertainty principle.*

This *uncertainty principle* means that if the measurement error $\varepsilon(Q)$ for position is reduced, the disturbance $\eta(P)$ in momentum increases, and conversely, if the disturbance $\eta(P)$ in momentum is reduced, the measurement error $\varepsilon(Q)$ for position increases.

In the same year, 1927, Kennard mathematically proved an inequality representing the quantum fluctuations (standard deviation) of position and momentum. Denoting the fluctuation of position as $\sigma(Q)$ and the fluctuation of momentum as $\sigma(P)$, the inequality is expressed as:

$$\sigma(Q)\sigma(P) \geq h/4\pi.$$

This is Kennard's inequality, which today is referred to as Kennard's *uncertainty relation.*

Heisenberg mistakenly believed that his *uncertainty principle*, which he anticipated from his thought experiment, was mathematically proven by Kennard's inequality. And thereafter, Heisenberg came to refer to

$$\varepsilon(Q)\eta(P) \geq h/4\pi$$

as *the uncertainty principle.*

As a result, Heisenberg's *uncertainty* regarding the error in position measurement, $\varepsilon(Q)$, and the disturbance in momentum, $\eta(P)$, came to be interpreted as a physical manifestation of the *uncertainty* expressed by Kennard's formula for the fluctuation in position $\sigma(Q)$ and momentum $\sigma(P)$. From a metaphysical perspective, one could interpret Kennard's inequality as a mathematical idea, a Platonic idea of existence, and Heisenberg's inequality, which appeared as the shadow of such an idea, as being recognized in the physical world.

In today's context, *the uncertainty relation* refers to Robertson's inequality which extends to the fluctuations of general physical quantities:

$$\sigma(A)\sigma(B) \geq 1/2|\langle[A, B]\rangle|.$$

In summary,

1. *Epistemological uncertainty* is *the uncertainty principle*, which indicates the limits of measurement in quantum systems.
2. *Ontological uncertainty* is *the uncertainty relation*, which indicates the inherent *uncertainty* of physical quantities.

It should be noted that in this treatise, *uncertainty* refers specifically to the mathematical *uncertainty relation.*

To explain the epistemological problem that one cannot accurately measure (determine) both the position and momentum of an object simultaneously, there is a need to provide a natural form based on an ontological approach. When the uncertainty in a position is multiplied by the uncertainty in momentum, the value is greater than 1.05/1027 g cm/s.

To measure (observe) an ur-presheaf p in S^, such p must be in an ur-particle state, requiring the specification of an object V in t-site S. In addition to specifying the ur-particle state over V, we also need another ur-particle state over $U \in Ob(S)$ to determine the rate of change of the states of an ur-entity. The larger l is in the following sequence of a factorization $V = V_0 \to V_1 \to V_2 \to \cdots \to V_l \to V_{l+1} = W$ of a morphism $V \to W$, the more detailed information is gained about the change of the ur-states between $p(V)$ and $p(W)$.

In the temporal topos theoretic method, for generalized *micro-time periods* or *ur-Planck objects* V and U, the uncertainty phenomenon is expressed as a *micromorphism* from V to U, i.e.,

$$V \to U. \tag{3.7}$$

See Sect. 2.2 on *Microcoverings* for the definition of a generalized micro-time period or ur-Planck objects. Expression (3.7) may be interpreted as the mathematical ontological view of our theory. As the epistemological view for an electron case, represented by an ur-presheaf e, the uncertainty phenomenon would be expressed as

$$e(V) \leftarrow e(U) \tag{3.8}$$

indicating the uncertainty in position between the ur-particle states $e(V)$ and $e(U)$ and the uncertainty of the rate of change $e(V) \leftarrow e(U)$.

The uncertainty principle implies that no particle is at rest at one place. If the position of a particle were to be determined, it would be impossible to define its momentum and vice versa.

3.6 The Measurement Phenomena and Many Universes Interpretation, i.e., Multiverse

When an ur-presheaf p is not in an ur-particle state, it makes no sense to ask where the particle is. The ur-presheaf p is *neither* everywhere *nor* nowhere. For example, when p is in an ur-particle state, by definition, there exists an object V in t-site S. Then the quantum state reduction occurs as $p(V)$ from the ur-wave state $p(-)$, where '−' indicates an object in S is unspecified.

Let us find the corresponding formulation in terms of t-site objects to the notion of a light cone. The ur-lightcone in t-site S with vertex V looks like this:

$$
\begin{array}{ccccc}
\cdots\cdots & \cdots & & \cdots & \cdots\cdots \\
\cdots & \nwarrow\uparrow\nearrow. & \cdots & & \cdots\cdots \\
\cdots & \cdots\; V_d\; \cdots & & \cdots & \ldots\; \text{level 3} \\
 & \nwarrow..\uparrow..\nearrow.\nwarrow\uparrow\nearrow..\nwarrow\uparrow\;\nearrow & & & \\
 & V_k \ldots\; V_i\; \ldots\; V_j & & & \text{level 2} \\
 & \nwarrow..\uparrow..\nearrow & & & \\
 & V. & & & \text{level 1}
\end{array}
\qquad (3.9)
$$

Let us assume that a t-site object V in diagram (3.9) corresponds to the present time. Among all the future choices which means all the t-linear morphisms from $\{V \to V_i\}_i$ from V, one particular object V_0 among all the possibilities $\{V \to V_i\}_i$ is specified for the actual future ur-particle state at V_0.

Herein lies *the measurement problem.* After the ur-particle state $p(V)$ of the present time, what causes a choice for a temporally linear morphism from V to the particular future-determining object V_0 among all the finitely many morphisms with vertex V as in (3.9)? Namely, we ask what causes the shift from the probabilistic ur-wave state $p = p(-)$ in the superposition to the collapsed state of $p(V_0)$, the ur-particle state for a *particular* object $V_0 \in Ob(\mathrm{S})$. We like to answer the question how the particular object V_0 is chosen from among the finitely many $\{V_i\}_i$ at level 2 with vertex V in diagram (3.9). The actually chosen V_0 must satisfy the commutativity of the diagram:

$$
\begin{array}{ccc}
\alpha & \to & \omega \\
\downarrow & \swarrow & \\
p(V_0). & &
\end{array}
\qquad (3.10)
$$

When the generalized-time period $V \in Ob(\mathrm{S})$ indicates the present time, all the temporally linear morphisms originating from V indicate all the possible futures. Before the present time determining object V, the particle was in superposition $p(-)$. Once again, after the present time, the particle is in an ur-wave state $p = p(-)$ at least for a Planck time period. For a future ur-particle state from the ur-particle state $p(V)$, there are finitely many possibilities, which is equal to the number of t-linear morphisms from V in the above diagram (3.9). Until the future ur-particle state is determined, p is in an ur-wave state superposition, which from now on we indicate as just p rather than $p(-)$. An ur-wave state p corresponds to the wave state described as $|\psi\rangle$, the temporal evolution of the deterministic quantity governed by a partial differential equation.

We will consider the following three cases:

Case (i): when $V_1 \neq V_0$ had been chosen within the ur-lightcone (3.9) with vertex V,
Case (ii): when V_2 is not in the ur-lightcone with vertex V, i.e., $V \to V_2$ is *not* a temporally linear morphism, the ur-particle state $p(V_2)$ can be within the same universe as $p(V_0)$, and.
Case (iii): when $p(V_3)$ is in a separate universe from our universe, $p(V_3)$ is not in the same universe as $p(V_0)$.

Note that the number of multiverses *equals* the number of temporally non-linear morphisms from V of case (iii).

If V_0 is the unique object in t-site S to satisfy the compatibility condition (3.10) with respect to an initial object α and a terminal object ω of t-topos S^, then we have provided a condition for the particular choice of the future by the uniqueness of V_0. Namely, the procedure of time from the present to the future is a deterministic choice of such a unique object V_0 so as to satisfy the commutative diagram (3.10).

If such a V_0 satisfying (3.10) is not unique, our temporal topos interpretation would be only a rephrasing in our language for measurement phenomena.

Summarizing the above: for a given moment at V, representing the present time, the future possibilities can be captured as all the t-linear morphisms from V to $\{V_\cdot\}$, where $\cdot = k, \ldots, 0, \ldots, i, \ldots, j$, on level 2 in the ur-light cone (3.9). Measurement phenomenon asks how and why the V is chosen, and how and why $\cdot = 0$ will be chosen as the immediate future of V, i.e., how the future is determined and reified.

We can consider an ur-lightcone with vertex V_l, including the microlevel for a covering $\{V \leftarrow V_l\}_l$ of V. As indicated above, *measurement phenomena relate to reality in macrocosm and in microcosm and time as a choice of history*, which are represented by morphisms in temporal topos S^ and temporal site S. In both microcosm and macrocosm, at each stage choices demanding the commutativity as in (3.10) are made in the transition from '*the now*' to the future.

Note As above let $q(V)$ be an ur-particle state of q at V. Let ε_j: $q_{j.} \to q = \prod q_j$ be a decomposition of q and let $\{V \leftarrow V_l\}_l$ be a covering of V. Then some of $\{q_j\}_j$ are in ur-particle states reified as $\{q_j(V_i)\}$ for a certain V_i. We regard *quantum fluctuations* as a combination of ur-particle states $\{q_j(V_i)\}$ and ur-wave states $\{q_k\}_k$ progressing according to the ur-lightcones.

3.7 The Issue of Determinism and Free Will

Free will in thoughts and actions as macro events originates from the activities of cells, molecules, and ultimately elementary particles. We capture and formulate the notion of *free will* viewed from the angle of measurement phenomena and ur-lightcone.

The issue of *free will* enters at V, corresponding to the present time, and at level 1 in the ur-lightcone (3.9).

If the uniqueness of the object V_0 at level 2 is guaranteed to make diagram (3.10) commutative, there is no chance for free will to exist. We will return to this topic in Chap. 5 as an issue of an initial object α of t-topos S^. See also [Kato-Nishimura (2020), (2021)] for related issues.

3.8 Gravitational Effect in Terms of T-Topos

We will give the t-topos theoretic interpretations of such phenomena as the gravitational effect, big bang, and even the pre-big bang.

We should ask what caused the big bang as a measurement phenomenon. We can ask "Why then?" and "Why not later?" We are assuming there was such an event as the big bang at the beginning of our universe. Were there any other choices besides our big bang? Dark energy and dark matter, so far without solid explanations, are serious conundrums.

Let P be a particle, possibly a subatomic particle. Let the associated ur-presheaf p with particle P be reified over V and $W \in Ob(\mathrm{S})$. The basic idea is as follows. Under more intense gravity, the ur-presheaf p is *less likely* to be in an ur-particle state. To be more precise; let

$$p(V) \to p(V_1) \to p(V_2) \to \cdots p(V_{k-1}) \to p(W) \tag{3.11}$$

be a microfactorization of $V \to W$ for p under no gravity. And let

$$p(V) \to p(V_1) \to p(V_2) \to \cdots \to p(V_{k^*-1}) \to p(W) \tag{3.11*}$$

be a microfactorization of $V \to W$ for p under non-zero gravity, where $V_0 = V$ and $V_k = V_{k^*} = W$. Then the *temporal topos gravitational hypothesis* (abbreviated as *t. g. hypothesis*) is an inequality

$$0 \leq k^* < k, \tag{3.12}$$

in the case where $k^* = 0$, i.e., $p(V_{-1})$, is interpretated as p being not in an ur-particle state. As an extremely intense gravitational case, where $k^* = 0$, particle P is always in a wave state, i.e., the associated ur-presheaf p is always in an ur-wave state. Then our interpretation of a pre-big bang is that the all objects of temporal topos S^ are in ur-wave state, and the beginning of our physical universe is when there exist ur-presheaves in an ur-particle state for the first time.

From the *t. g. hypothesis*, i.e., inequality (3.12) we get the following consequences:

(1) We consider two particles P and P^*. Let us assume that two ur-particles p and p^* associated with particles P and P^* are ur-entangled. Our thought experiment is the following. Particle P^* falls into a black hole, and the other particle P stays outside the black hole. When P^* is close enough to the center of the black hole, the corresponding ur-presheaf p^* cannot be in an ur-particle state. Then the ur-entangled ur-presheaf p with p^* is not allowed to be in an ur-particle state, i.e., p^* is in an ur-wave state, and particle P must be also.

(2) Pre-big bang may be expressed as not a single ur-presheaf is reified. Namely, none of the objects in t-topos S^ is in an ur-particle state; i.e., all the ur-presheaves are in an ur-wave state. As for an initial object α and a terminal object

ω, they were in neither an ur-particle state nor in an ur-wave state. Rather, no words may better suit α and ω, since they are both constant functors.

Note There are many books and articles available on quantum physics and quantum gravity which are written by authors such as Roger Penrose, Christopher Isham, Julian Barbour, Brain Greene, Lee Smolin, David Deutsch, et alia.

3.9 Synopsis

After a brief review on the quantum world of microcosm phenomena, the universal and structural physical aspects of the observer and the observed are explained in terms of categories and sheaves. Then, we focus on quantum strangeness: the particle-wave duality, quantum entanglement, quantum uncertainty, and measurement phenomenon. As an example, we give a gravity formula showing that the mass of a particle affecting spacetime, which is measured by the observer, is independent of the distance from the particle to spacetime as it should be. Then the constancy of an ur-functor of spacetime plays an effective role for the formulation of gravity. We also discuss applications of temporal topos gravitational hypothesis, abbreviated as t.g. hypothesis, to the big bang notion. See [Barbour (1999)], [Butterfield-Isham (1999), (2001)]. Also compare our approach with [Guts-Grinkevich (1996)] and the abstract differential geometric approaches as in [Mallios-Zafiris (2016)], [Mallios-Raptis (2003)] and [Raptis (2006, 2024)].

Chapter 4
Cognitive and Biological Sciences

Contents

Roger Penrose writes,

> Perhaps what we mainly need is some subtle change in perspective ——— something that we all have missed ....
>
> p. 1045 of The Road to Reality

Namely, Penrose also seems to point us to:

> Thus, the task is not so much to see what no one has yet seen, but to think what nobody yet has thought about that which everyone sees.
>
> Arthur Schopenhauer in his two-volume work *Parerga und Paralipomena*

The goal of this chapter is to give a sheaf theoretic (i.e., temporal topos theoretic) interpretation of cognitive phenomena. As an example:

> The brain is responsible for our thoughts: consciousness which is sometimes followed by physical actions. However, each cell, for that matter, each atom alone cannot function as a conscious entity. We need to explain explicitly with mathematical rigor how a global entity consisting of a huge number of trivial entities like atoms becomes capable of having thoughts. First, we need to explain the mechanisms leading to how such a collection of a large number of trivial entities can animate a global object as a living and thinking creature.

As already noted, we will consider a specific category S^ of (ur-)presheaves whose codomain (target) category O has an initial object and a terminal object. Such a category S^ must satisfy the temporal topos theoretic constraints as listed in Chap. 2. One more element of our theory is the role played by sheaf theory, i.e., the concept of descent. In other words, we need a 'gluing device' connecting local data consisting of trivial data to obtain a non-trivial global object.

G. C. Kato, *Temporal Topos Methods for the Philosophy of Natural Sciences*,
https://doi.org/10.1007/978-981-96-2420-1_4

As has already been mentioned in Chap. 1, our theory consists of only the following three categories:

(i) Category S^, temporal topos which is a category representing all the entities from the inanimate to the animate.
(ii) Category S, temporal site which is a category playing a more fundamental time concept for objects of S^.
(iii) Category O, the codomain category for S^ which has an initial object and a terminal object.

There are deeper phenomena still challenging our theory. Yet, we hope that our methods will perform as a harbinger for more accurate and detailed future study in all fields touched upon in this treatise.

The level of difficulty in this century is to answer such questions in biology as how a DNA sequence is responsible not only for a global shape, including embryology, but also for medical information. This may be even much more profound than answering such a question asked many centuries ago in physics as why a planet gravitationally can affect another planet without touching it.

In biology, we need chemistry laws for molecular behaviors to answer a fundamental question such as 'What does it mean for a physical entity to be alive?' The laws in chemistry, on the other hand, can be explained in principle by quantum physics which is interpreted in terms of mathematical concepts.

According to Roger Penrose, there are three worlds: the mathematical, the biological, and the physical. In his book *The Road to Reality*, Penrose refers to these three worlds as the platonic mathematical world, the mental world, and the physical world, respectively.

4.1 The Penrose Triangle

Since not all physical matter is associated with life, we have the embedding:

$$\text{the biological} \subset \text{the physical}. \tag{4.1}$$

Since not every mathematical concept is relevant to the physical realm, we have the following embedding:

$$\text{the physical} \subset \text{the mathematical}. \tag{4.2}$$

However, every mathematical activity is part of brain activity, so finally:

$$\text{the mathematical} \subset \text{the biological}. \tag{4.3}$$

In view of philosophical materialism, inclusions in (4.1) and (4.3) may be reasonable. However, the inclusion in (4.2) is not straightforward. Note that Penrose indicates these relationships (e.g., *Shadow of the Mind*, p. 414, in [Penrose (1994)]), by an epimorphic (surjective) arrow instead of an embedding inclusion notation as above. For example, (4.3) 'the mathematical $\subset$ the biological' becomes the surjective (onto) mapping:

$$\{\text{a part of the biological world}\} \rightarrow \{\text{the entire mathematical world}\}.$$

In our treatise, we provide a solution to this enigma by introducing an initial object and a terminal object of a temporal topos.

Here is the basic approach: by expressing a procedure of 'gluing' local data $\{u_i$ over $X_i \in Ob(\mathrm{S})\}$ through either an equality or an isomorphism over the 'overlaps' of local data, we can obtain a global object (u, X). Once such a global object u over $X \in Ob(\mathrm{S})$ is obtained, we need to confirm that the restriction of u to each X_i coincides with the originally given u_i over each X_i.

Under an existence hypothesis of an initial object α and a terminal object ω of a t-topos S^, together with the notion of an ur-lightcone in a t-site S, we glue local data to obtain a global object. Note that an initial object α and a terminal object ω of a t-topos S^ are constant functors. Hence, the '*values*' in the codomain category O of constant functors α and ω are an initial object A_O of O for α and a terminal object Z_O of O for ω for any abject (i.e., *variable*) V in S, respectively.

One of the purposes of this chapter is to show that an initial object in t-topos is capable of explaining the cognitive behaviors of conscious entities.

We need to go back to one of the original motivations for our approach in terms of topos and sheaf-descent theory. That is, all the ur-presheaves associated with subatomic particles to global objects like a stone, an apple, a human being, a star, etc., are temporally parameterized by objects of a t-site S. We must formulate explicitly how they are self- organized so that a global entity can emerge from local entities. Atoms are compounds of elementary particles; molecules are compounds of atoms. A final destiny as a global object, for example, is an entity with life, i.e., a living organism.

In the reverse direction, i.e., decomposing a global object into smaller pieces, we express the process from the large to the small in terms of presheaves as follows.

Let q be an ur-presheaf corresponding with a global object, and then express $q = \coprod q_k$ as a co-product of $\{q_k\}_k$ for a decomposition of q into smaller ur-subpresheaves $\{q_k\}_k$ of q. Suppose q is reified as $q(V)$ at $V \in Ob(\mathrm{S})$ and let $\{V \leftarrow V_j\}_j$ be a covering of V. See Appendix 2 for the definition of a covering and the notion of a co-product. Then assume that all the ur-subpresheaves of q are reified at some of those $\{V_j\}_j$ in the covering of V. That is, covering $\{V_j\}_j$ has enough objects for the decomposition $\coprod q_k$.

Our methods and approaches toward life sciences are based on "gluing" locally given data to get a unique morphism δ^q from an initial object α to q. For an object q of t-topos S^ over V in S, we have a unique morphism $\delta^q(V)$ from an initial object α, which is associated with the cognitive state of q over the generalized time-period V

in the t-site S. We can regard an object V of S as a parameter for an ur-particle state of q.

As natural phenomena, we already have globally emerged objects. We have been analyzing and studying the local objects through the scientific methods by decomposing such global objects. Hence, our task is to explain the mechanisms for the phenomena arriving at an emerging global object from the local data.

Following is a definition of consciousness based on temporal topos theory.

Definition Let q be an ur-presheaf, i.e., an object of t-topos S^, representing a conscious entity. Then *consciousness of* q, i.e., *cognitive ur-wave state of entity* q is a uniquely determined morphism from an initial object α to q which we denote as.

$$\delta^q. \tag{4.4}$$

More explicitly, we can define a pair (q, δ^q) as the consciousness of q.

Therefore, we define *the concept of consciousness itself* as

$$\delta. \tag{4.5}$$

without the specification of an object in S^.

Remarks

(1) Returning to the definition of δ: a natural transformation δ between functors, i.e., a morphism of functors, does not exist without an object q of S^. This is because the uniquely determined morphism δ^q from α to q exists (i.e., is reified at q) only when an object q is specified. Namely, without the specification of an object q, δ does not exist as a morphism of ur-functors from an initial object α to q. That is, δ alone is only a concept with the potential to exist until an object in S^ is chosen. This remark is relevant to philosophical aspects of temporal topos. See Chap. 5 on philosophical roles played by temporal topos.

(2) *For mathematically inclined readers*, one can regard δ as a correspondence from objects in S^ to the natural transformations from an initial object of S^ to objects in S^, i.e., assigning δ^q for each $q \in Ob(\text{S}^\wedge)$. Namely, we can consider δ as a mapping δ: $Ob(\text{S}^\wedge) \to \{\text{morphisms of ur-functors}\}$, i.e., $q \in Ob(\text{S}^\wedge) \to \delta^q$.

(3) As a categorical observation, if an initial object α is the only object in S^, δ^α must be equal to the identity morphism $id_\alpha : \alpha \to \alpha$, i.e., $\delta^\alpha = id_\alpha$.

Recall that since t-site S plays the role of a parameter category of cognitive ur-states $\delta^q(V)$ of q over V in S, we can express when the consciousness of q takes place. In other words, such a conscious ur-state of q is parameterized by t-site objects $\{V\}$ for $V \in Ob(\text{S})$. Then a pair $(q(V), \delta^q(V))$ of object $q(V)$ and $\delta^q(V) : \alpha \to q(V)$ indicates the conscious ur-particle state of q during the generalized-time period V indicating 'when' such a conscious ur-state of q is taking place, also written as:

$$\begin{array}{c} \alpha \\ \downarrow \delta^q(V) \\ q(V). \end{array} \tag{4.6}$$

As noted above, this definition of the consciousness of q can be temporally parameterized in terms of objects in t-site S. Namely, for the consciousness (q, δ^q) of q, we can formulate the physical aspect of 'when and where' for the ur-state $q(V)$ through the morphism $\sigma_q(V)$ from $q(V)$ to ω, and the consciousness of q occurs as a pair $(q(V), \delta^q(V))$ for $V \in Ob(\mathrm{S})$. That is, the consciousness of q has a 'parameter' in S; an ur-particle state can be temporally expressed for the past, the present, and the future. That is, a sequence of temporally linear morphisms $U \to V \to W$ among three objects U, V, and W of t-site S, represents the past, the present, and the future, respectively. Object V corresponds to the usual present time t_0 (i.e., in terms of time associated ur-presheaf τ, $\tau(V)$ is corresponding to t_0). Then we get the canonically induced sequence $q(W) \to q(V) \to q(U)$ whose cognitive morphisms are given by $\delta^q(W)$, $\delta^q(V)$, and $\delta^q(U)$, respectively.

Summarizing the above: the consciousness of q over the present time object V of S is a pair $(q(V), \delta^q(V))$, where the physical position and time of q at V is the pair $(q(V), \sigma_q(V))$ where morphism $\sigma_q(V)\colon q(V) \to \omega = \omega(\text{-}) = \omega(V) = (\tau(V), \kappa(V)) = (\tau, \kappa)$. Namely, a more appropriate description of *the consciousness* $(q(V), \delta^q(V))$ *of* q *over* V may be captured as the following commutative diagram, where the morphism $\sigma_q(V)$ indicates the physical ur-state of $q(V)$:

$$\begin{array}{ccc} \alpha & \to & \omega \\ {\scriptstyle \delta^q(V)} \searrow & & \nearrow {\scriptstyle \sigma_q(V)} \\ & q(V). & \end{array} \tag{4.7}$$

The difference between a conscious entity and a non-conscious entity is the range of choices of t-linear morphisms from V. Let $h\colon V \to W$ be a t-linear morphism. Since q is a contravariant functor, there is a canonically (functorially) induced morphism:

$$q(h) : q(W) \to q(V), \tag{4.8}$$

for any object q in S^ whether q is a conscious entity or not. In the above expression the range of choices has the following meaning. The number of t-linear morphisms from V is *fewer* for the case where q is a conscious entity because of q's *will* compared with a non-conscious entity.

Let us review: for the canonical morphism $q(h)\colon q(W) \to q(V)$ in (4.8), we have the following triangular commutative diagram:

$$\begin{array}{ccc} & \alpha & \\ \swarrow & & \searrow \\ q(W) & \to & q(V), \end{array} \tag{4.9}$$

I.e., $\delta^q(V) = q(h) \circ \delta^q(W)$, and diagram:

$$\begin{array}{ccc} & \omega & \\ \nearrow & & \nwarrow \\ q(W) & \rightarrow & q(V) \end{array} \tag{Dual 4.9}$$

is commutative, i.e., $\sigma^q(W) = \sigma^q(V) \circ q(h)$.

In addition, as the cognitive aspect of q, we consider a non-canonical morphism called a *relevant mnemonic morphism* $j_V{}^W$: $q(V) \rightarrow q(W)$ so that the following diagram is commutative.

$$\begin{array}{ccc} \alpha & \rightarrow & \omega \\ {}^{\delta^q(V)}\downarrow & & \uparrow_{\sigma_q(W)} \\ q(V) & \rightarrow & q(W). \end{array} \tag{4.10*}$$

In diagram (4.10*), the diagonal morphisms $\delta^q(W)$: $\alpha \rightarrow q(W)$ and $\sigma_q(V)$: $q(V) \rightarrow \omega$ are omitted. See Sect. 4.2 below for the commutativity of the triangular diagram parts.

Remarks We have the following three equivalent expressions (1), (2), and (3):

(1) the relevancy of $\delta^q(W)$ with respect to $\delta^q(V)$ in (4.10*), meaning that the ur-cognitive state $\delta^q(W)$ of q over W and the ur-cognitive state $\delta^q(V)$ of q over V are coherently related,

(2) the commutativity $\delta^q(W) = j_V{}^W \circ \delta^q(V)$ of diagram

$$\begin{array}{ccc} \alpha & & \\ {}^{\delta^q(V)}\downarrow & \searrow^{\delta^q(W)} & \\ q(V) & \rightarrow & q(W), \end{array}$$

holds, and

(3) the mnemonic morphism $j_V{}^W$ is compatible with $\delta^q(W)$ and $\delta^q(V)$.

4.2 The Animate and the Inanimate

Let q be an arbitrary ur-presheaf in microcosm or macrocosm. We shall review first the above basic facts. Since q is a contravariant ur-presheaf for a morphism h: $V \rightarrow W$ in S, there is a canonically induced morphism $q(h)$ (i.e., *functorially induced*) from $q(W)$ to $q(V)$, which we write as $q(h)$: $q(W) \rightarrow q(V)$ as in (4.8), satisfying the commutativity in (4.9). Then a mnemonic morphism $j_V{}^W$, i.e., a *non-canonical* morphism, is induced (or activated) by α through δ. The direction of such a non-canonically induced morphism is the opposite direction of the functorially induced morphism as in (4.8). Namely, we express a non-canonical morphism as follows:

$$j_V^W : q(V) \rightarrow q(W). \tag{4.11}$$

We will focus on the left-half of (4.10*). If the following diagram is commutative, i.e., satisfying $\delta_q(W) = j_V{}^W \circ \delta_q(V)$:

$$\begin{array}{ccc} & \alpha & \\ \swarrow & & \searrow \\ q(V) & \rightarrow & q(W), \end{array} \tag{4.12}$$

Then a non-canonical morphism $j_V{}^W$: $q(V) \rightarrow q(W)$ is said to be a *relevant mnemonic morphism for q and f*: $V \rightarrow W$ *with respect to* $\delta_q(W)$ and $\delta_q V)$. We can also say that $\delta^q(W)$ and $\delta^q(V)$ are relevant to the non-canonical morphism $j_V{}^W$: $q(V) \rightarrow q(W)$. Our formulation for a transition from inanimate particles to life can be phrased as follows. Consider a decomposition $\coprod q_k$ of a conscious entity q, and if necessary, a further decomposition of q_k can be considered. Ultimately, we arrive at a microdecomposition of q as discussed in Chap. 2. In the opposite ascending direction (i.e., beginning from a microdecomposition), at a certain level of decomposition of q, which we denote as q_{an}, ur-presheaf q_{an} begins to obtain the non-canonical (non-functorial) morphism. Then q_{an} is said to be an *animated ur-presheaf* when the following diagram becomes commutative:

$$\begin{array}{ccc} & \alpha & \\ \swarrow & & \searrow \\ q_{an}(V) & \rightarrow & q_{an}(W). \end{array} \tag{4.12an}$$

Note-Definition Commutative diagram (4.12) is corresponding to the left-half of (4.10*), and the right-half

$$\begin{array}{ccc} & \omega & \\ \nearrow & & \nwarrow \\ q(V) & \rightarrow & q(W) \end{array} \tag{4.13}$$

is the inanimate aspect of (4.10*), which is the definition of a *physical aspect of mnemonic morphism* $q(V) \rightarrow q(W)$.

Remark An ur-presheaf q is *animate* if diagram (4.12) is commutative for a relevant mnemonic morphism. On the other hand, a ur-presheaf q is *inanimate* if diagram (4.13) alone is a commutative diagram, which can be called a *physically relevant mnemonic morphism*. One could call it an *artificial-intelligence relevant mnemonic morphism* rather than a physically relevant mnemonic morphism.

Note For the canonical case, namely, we have functorially $q(W) \rightarrow q(V)$ for a t-linear morphism $V \rightarrow W$. Then we always have commutative diagrams as follows for α and ω, respectively.

$$\begin{array}{ccc} & \alpha & \\ \swarrow & & \searrow \\ q(W) & \rightarrow & q(V) \end{array}$$

and

$$\begin{array}{ccc} & \omega & \\ \nearrow & & \nwarrow \\ q(W) & \rightarrow & q(V). \end{array}$$

An observation morphism (for this case, we have $p(W) \rightarrow q(V)$, $p \neq q$) and a mnemonic morphism are *not* induced functorially, which are examples of a non-canonical morphism.

First let us begin with a general case. Let q be an observer and let p be the observed by q. Then q can be a conscious entity, but the observed object p does not have to be a conscious entity. An observation may be considered as a measurement of p by q, or an information flow from p to q as we have mentioned already. If we need to indicate the time concept for each observation, we need a parameter indicating when such an observation takes place. We indicate a non-canonical observation morphism with parameters from ur-presheaf p to ur-presheaf q as

$$p(V) \rightarrow q(W). \tag{4.14}$$

Then the temporally linear morphism $V \rightarrow W$, appearing in (4.14), means that V precedes W in the temporal site sense, i.e., V corresponding to, e.g., a past time $\tau(V)$ and W corresponding the present time $\tau(W)$.

Next, we consider the case when $p = q$. Then such a self-observation may be considered as a relevant mnemonic morphism of the past period V from the present W. We have

$$\begin{array}{ccc} & \alpha & \\ \swarrow & & \searrow \\ q(V) & \rightarrow & q(W). \end{array} \tag{4.15}$$

Our fundamental and motivational question is the following. Even though subatomic particles constituting animate entities, and as inanimate subatomic particles are identical, i.e., isomorphic as ur-presheaves, their constituted global objects can be completely different. Our interpretation via temporal topos theory is that each organizing process through the 'gluing' local data differs. To formulate the connection between the wholeness as a nontrivial global object and mutually communicating trivial local parts, the notion of mnemonic morphism plays a significant role.

Definition For a t-linear morphism h: $V \rightarrow W$, if a relevant mnemonic morphism ${j^V}_W{:}q(V) \rightarrow q(W)$ exists satisfying the commutativity $\delta^q(W) = {j_V}^W \circ {}^q(V)$ in (4.15), the ur-presheaf q represents a cognitively *animate entity*, i.e., a living thing during the period from V to W. Otherwise, if only the canonical morphism $q(h)$: $q(W) \rightarrow q(V)$ exists satisfying the compatibility commutative diagrams (4.9) and (Dual 4.9), then such an ur-presheaf represents an *inanimate entity* during $V \rightarrow W$.

We indicate the following for non-canonical morphisms:

a unique morphism δ^q from an initial object α induces non-canonical morphisms,

which we expressed by saying '*influenced* (*affected*)' by α, i.e., by δ* in the above sentence, where superscript * of δ indicates an arbitrary object of S^. This very matter is related to the notion of *free will* and *measurement phenomena* in the sense that the number of choices from V, which is representing, e.g., the present to the future, is fewer in the ur-lightcone with vertex V in (3.9) in Chap. 3, which is:

$$\begin{array}{cccccc} \cdots\cdots & \nwarrow\uparrow\nearrow. & \cdots\cdot & \cdots\cdots & & \\ \cdot\cdot\ \ \cdots\cdot & V_d\ \cdots & \cdots & \cdot\cdot\ . & & \text{level 3} \\ \multicolumn{4}{c}{\nwarrow..\uparrow..\nearrow.\nwarrow\uparrow\nearrow..\nwarrow\uparrow\ \nearrow} & & \\ \multicolumn{4}{c}{V_k \ldots\ V_i\ \ldots\ V_j} & & \text{level 2} \\ \multicolumn{4}{c}{\nwarrow..\uparrow..\nearrow} & & \\ \multicolumn{4}{c}{V.} & & \text{level 1} \end{array}$$

Experience-Observation Suppose that you were unsuccessfully trying to recall a memory from your past. Then you may have the following experience.

(i) There was a moment when you got 'it' but did not know what 'it' exactly was yet.
(ii) Immediately after that moment, you knew what 'it' was.

There is a time gap between (*i*) and (*ii*). The above example indicates that δ* induces a mnemonic morphism.

A formulation for this phenomenon of a brain function in terms of temporal topos language becomes as follows. The very moment when you felt you got 'it' is expressed as $\delta^q(W)$. The memory from the past of what 'it' exactly was is expressed as the mnemonic morphism $j^V{}_W$ from the past ur-state over V to the present ur-state over W. The very issue comes to mind as $j^V{}_W \circ \delta^q(V)$ at stage (*ii*) satisfying the commutativity

$$\delta^q(W) = j^V{}_W \circ \delta^q(V)$$

in diagram

$$\begin{array}{ccc} & \alpha & \\ \delta^q(V) \swarrow & & \searrow \delta^q(W) \\ q(V) & \rightarrow & q(W). \end{array}$$

4.3 Applications

We scientists have experienced several *eureka*-moments. For example, a mathematician tried to solve a problem for weeks, months, or even years. Suppose that a certain amount of time was spent solving a problem without success. Namely, in spite of effort, we gain only independent and unrelated information for a complete solution to the problem, perhaps sometimes with some success making a few connections to previous efforts. After then, the moment came to recognize a complete solution to the problem. Then previous efforts are all connected and coherent to the complete solution in the sense that previous local pieces of understanding were relevant parts of the complete solution.

On the other hand, let us consider the completely different world of insects, especially the well-organized society of leaf-cutting ants. A colony of ants even knows how to ferment leaves for food. However, an individual ant does not seem to have knowledge to make fermented leaves. However, globally, as a globally emerged colony consisting of individual ants, their behavior is highly organized.

In the above, we expressed the change from the first topic on discovery of a solution to another topic on an ant colony as a completely different world. Both topics share a procedure of moving from the local trivial to an organized global object.

We divide the phenomena into three groups: Type I, Type II, and Type III.

Type I

We will give a more precise and explicit formulation than the above for the moment of a discovery, i.e., the eureka-moment, or simply a realization moment of recovered thoughts. In terms of temporal topos language, we formulate the following: non-coherent, more or less discrete individual information data are (suddenly) reified as a global information as a eureka-moment. Furthermore, the restriction of the obtained global object to each previous local datum matches the experienced data.

Consider a factorization of a temporally linear morphism $V \to W$ as

$$V = V_{n+1} \to V_n \to V_{n-1} \to \ldots \to V_1 \to V_0 = W \tag{4.16}$$

by t-linear morphisms $\{h_i\}_i$ and objects $\{V_i\}_i$ in a t-site S, i.e., h_i: $V_i \to V_{i\text{-}1}$, $i = 1, 2, \ldots, n + 1$. Then, there is a canonically induced sequence by an arbitrary object q in S^, that is,

$$q(V) \leftarrow q(V_n) \leftarrow q(V_{n-1}) \leftarrow \cdots \cdot \leftarrow q(V_0) = q(W), \tag{4.17}$$

where for $q(h_i)$: $q(V_{i\text{-}1}) \to q(V_i)$.

Our cognitive states (e.g., thoughts) generally vary as time passes. We can focus on a certain issue hourly, daily, or even monthly. To express such focused thoughts on one issue or topic, we have introduced a relevant mnemonic morphism

$$q(V) \to q(W),$$

which satisfies the following commutative diagram with respect to an initial object α of t-topos S^

$$\begin{array}{ccc} & \alpha & \\ \swarrow & & \searrow \\ q(V) & \rightarrow & q(W). \end{array}$$

Let us consider the following more general commutative diagram based on the above review comment consisting of relevant mnemonic morphisms:

$$\begin{array}{c} \alpha \\ \cdots \swarrow \quad \downarrow \quad \searrow \quad \cdots \\ q(V) \rightarrow q(V_n) \rightarrow q(V_{n-1}) \rightarrow \cdots \rightarrow q(V_0) = q(W) \end{array} \tag{4.18}$$

where we use the same notation for mnemonic morphisms as before,

$$j_{V_{i-1}}^{V_i} : q(V_i) \rightarrow q(V_{i-1}), i = 1, 2, \ldots, n-1.$$

On the other hand, the canonically induced morphism is given by

$$q(h_i) : q(V_{i-1}) \rightarrow q(V_i),$$

where h_i: $V_i \rightarrow V_{i-1}$.

Next, we will give a formulation for a eureka-moment. First, we need a definition for recalling thoughts in the past.

Definition For a factorization $V \rightarrow V_n \rightarrow V_{n-1} \rightarrow \cdots \rightarrow V_1 \rightarrow V_0 = W$ in (4.16) and (4.17), we define the *i-fold restriction* of δ(*W*) by

$$\delta(V_i) = q(h_i) \circ q(h_{i-1}) \circ \cdots \circ q(h_1) \circ \delta(W). \tag{4.19}$$

Notice that Eq. (4.19) can be rewritten also as

$$\delta(V_i) = q(h_1 \circ h_2 \circ \cdots \circ h_i) \circ \delta(W)$$

because of the contravariant-ness of ur-functor *q*. (See the definition of a contravariant functor in Appendix I). Then, we say that δ(*W*) *restricts well* when we have *i-fold restrictions* of δ(*W*) for all $i = 1, 2, \ldots, n\text{-}1$, for the factorization shown in (4.16).

Then we need to provide a formulation expressing *W* corresponding to such a eureka-moment, i.e., showing that a new insight is not obtained after *W*.

Definition Let *h*: $W \rightarrow U$ be any t-linear morphism. If we have an equality between the restrictions:

$$\delta(V_i) = q(h_i) \circ q(h_{i-1}) \circ \cdots \circ q(h) \circ \delta(U) = q(h_i) \circ q(h_{i-1}) \circ \cdots \circ \delta(W)$$

i.e., indicating the stability of the restrictions after W, then W is said to be a *first Eur-object* of S for the sequence (4.18).

Though the following assertion is a highly technical matter, we need to confirm it.

Technical Remark *for mathematically inclined readers*

For a sequence of t-linear morphisms

$$\to V_{l+1} \to V_l \to \cdots\cdot V_1 \to V_0 = W, \tag{4.20}$$

where h_l: $V_l \to V_{l\text{-}1}$, we define $V_l \times_W V_j = V_{l+j}$. Then we assert that

$$\{W \leftarrow V_l\}_l$$

is a covering of W in S.

We will provide a proof for this assertion as follows.

To show $\{W \leftarrow V_l\}_l$ is a covering of W, we must confirm three conditions (s.1), (s.2), and (s.3) for the definition of a covering as a part of a Grothendieck topology requirement as in Appendix II. We will demonstrate (s.2); for (s.1) and (s.3), see Remarks below.

Notice first that for an arbitrary t-linear morphism $U \to W$, there exists V_n for a certain integer n so that $V_n = U$ holds. Then the following commutative diagram

$$\begin{array}{ccc} U \times_W V_i & \to & V_i \\ \downarrow & & \downarrow \\ U & \to & W \end{array} \tag{4.21}$$

becomes

$$\begin{array}{ccc} V_{n+i} = V_n \times_W V_i & \to & V_i \\ \downarrow & & \downarrow \\ V_n & \to & W. \end{array} \tag{4.22}$$

Then for any $r \geq n + i$, consider the following diagram satisfying the usual universal mapping property for a fiber product $U \times_W V_i$ over W.

$$\begin{array}{cccc} V_r & & & \\ & \searrow & & \\ & V_{n+i} & \to & V_i \\ & \downarrow & & \downarrow \\ & V_n & \to & W. \end{array} \tag{4.23}$$

More explicitly, morphisms, i.e., the horizontal morphism $V_{n+i} \to V_i$ and the vertical morphism $V_{n+i} \to V_n$ are given by

$$h_{i-1} \circ h_{i-2} \circ \ldots \circ h_{n+i},$$

and

$$h_{n-1} \circ h_{n-2} \circ \ldots \circ h_{n+i},$$

respectively, and the tilled morphisms $V_r \to V_i$ and $V_r \to V_n$ are given by

$$h_{n-1} \circ h_{n-2} \circ \ldots \circ h_{n+i},$$

and

$$h_{n-1} \circ h_{n-2} \circ \ldots h_r,$$

respectively. Then there exists a unique morphism $V_r \to V_{n+i}$

$$h_{n+i-1} \circ h_{n+i-2} \circ \ldots h_r,$$

satisfying the universal mapping property. Namely, for the fiber product over W, i.e., the composition of the unique morphism $h_{n+i-1} \circ h_{n+i-2} \circ \cdots \circ h_r$ with the horizontal morphism $h_{i-1} \circ h_{i-2} \circ \cdots \circ h_{n+i}$ is the tilled morphism $h_{i-1} \circ h_{i-2} \circ \cdots \circ h_r$, and the composition of the unique morphism $h_{n+i-1} \circ h_{n+i-2} \circ \cdots \circ h_r$ with the vertical morphism $h_{n-1} \circ h_{n-2} \circ \cdots \circ h_{n+i}$ is the tilled morphism $h_{n-1} \circ h_{n-2} \circ \cdots \circ h_r$.

Consequently, $\{W \leftarrow V_l\}_l$ is a covering of W in S.

Remarks

(1) The requirement (s.1), i.e., the identity morphism $id_W\colon W \to W$, is a t-linear morphism. As for (s.3), for covering $\{V_l \leftarrow V_{lk}\}_k$ of V_l, the composite $\{W \leftarrow V_{lk}\}_{l,\,k}$ is a covering of W consisting of t-linear morphisms.

Many scientists have experienced puzzlingly sudden moments of scientific discoveries. Well-known scientists who described their experiences of scientific discoveries in their essays are Henri Poincaré, Jacques Hadamard, Herman Weyl, and Kiyoshi Oka. Such a *first Eur-object* W of t-site S for the sequence $V \to V_n \to V_{n-1} \to \cdots \to V_1 \to W$, with respect to q, corresponds with such a discovery moment. We quote "… I still remember vividly the night in my Göttingen Studentenbude when this idea came to me, seemingly without effort, and how greatly surprised I was that it actually worked. In the meantime, the kerosene lamp on my desk had started to smoke and the soot came down in the flakes from the ceiling." (Hermann Weyl's *Mind and Nature, Selected Writings on Philosophy, Mathematics and Physics*, Princeton University Press, 2009, page 164.) See also [Hadamard (1945)].

(2) Type I formulation works for the following example. If you, as an observer q, are allowed to see a large painting from a distance of a few centimeters away from the painting, then you would not grasp the global image of the painting. If you

are allowed to move around to see the painting, still keeping the same distance from the painting, and also if you are capable of gluing the information on locally observed images of the painting, you may be able to guess the global image of the painting. The local observation data are coming from a sequence of t-linear morphisms $V_1 \to V_2 \to \cdots \to W$ as you move around the painting keeping the short and same distance from the painting. As a global view of the painting, you would be able to recognize the painting. If you prefer to formulate from a different view, you need to consider $\{p(V_i)\}_i$ with the commutative diagram:

$$\begin{array}{ccc} & \alpha & \\ {}^{\delta^p(V_i)}\swarrow & & \searrow{}^{\delta^q(V_i)} \\ p(V_i) & \to & q(V_i). \end{array}$$

(3) In case of a music piece b, e.g., by Bach, if you, a listener q, hear the first note, the second note, …, after several notes you will be able to recognize the piece, provided that you already know the piece. You can consider the corresponding sequence $V_1 \to V_2 \to \cdots \to W$ of t-linear morphisms for progression of time listening to those notes.

(4) Judging *only and simply* by the categorial roles played by an initial object α of t-topos, we call such an initial object α as the cognizance source, the source of perception, thing-in-itself (Ding an sich) in the sense of Kant, or even *Reality*. More on theological aspects of Eastern thoughts: α corresponds to *Brahman*, and δ corresponds to *Atman*. See Chap. 5 for more about the role of an initial object α of temporal topos in philosophy.

Type II

This type of application is for a phenomenon with a unified whole consisting of individuals. As we will see shortly in Type II, our focus is on the structures in t-topos S^, which is less sensitive to a t-site S. That is, for such phenomena, we can give formulation without involving t-site objects. Technically speaking, such an application is a method to consider a topos as a site, where a direct sum of presheaves plays the role of a covering. Having said that, a truly accurate treatment is the combination of Type I and Type II. That is, we can consider several objects for Type I, i.e., time sensitive (or time dependent) and Type II, an organized whole as a direct sum of presheaves.

Categorically speaking, we consider

$$\begin{array}{c} S^{\sim} \\ \downarrow \pi \\ S^{\wedge}, \end{array} \tag{4.24}$$

where an object of S˜ is

$$\begin{array}{c} \alpha \\ \downarrow \delta^{q} \\ q \end{array} \tag{4.25}$$

So that the projection $\pi(\delta^q) = q$.

As an example, let us consider an organized colony of ants. From each individual behavior alone, we cannot grasp the behavior of the well-organized colony. In this case, each individual ant is the local, and its colony is the global. Let q be the ur-presheaf associated with a whole colony and let q_j be individual ants. Note that we can express the relationship between the colony and individual ants as the direct sum of q_j, where

$$\varepsilon_j : q_j \rightarrow \coprod q_j = q \tag{4.26}$$

is called the *ith coprojection morphism* from q_j to $\coprod q_j$. In the diagram below, for the coprojection ε_j: $q_j \rightarrow q = \coprod q_j$, we have the following commutative diagram

$$\begin{array}{ccc} & \alpha & \\ & \downarrow & \searrow \\ \coprod q_j = & q & \leftarrow \; q_j . \end{array} \tag{4.27}$$

Then note that for an observer p of the whole colony q as a global object, i.e., s: $q \rightarrow p$, we have information about the individual ant q_j, which is expressed in the diagram

$$\begin{array}{ccc} q & \leftarrow & q_j \\ \downarrow & & \\ p & & \end{array} \tag{4.28}$$

by composing the co-projection morphism ε_j with the observation morphism from q_j to p. That is, understanding the behavior of a global colony, we know the why and how of an individual ant's behavior.

On the other hand, as we have explained in 1.5 of Chap. 1, we will show that the incompatibility of a morphism composition is responsible for not knowing the whole from local observations.

When p observes each q_j, we have the following diagram

$$\begin{array}{ccc} q & \leftarrow & q_j \\ & & \downarrow \\ & & p. \end{array} \tag{4.29}$$

We cannot grasp the entire colony's global behavior by observing an individual q_j. Namely, discrete independent information on an individual ant can provide no (global) information about the colony as a whole. This is because an observation morphism $q_j \rightarrow p$ cannot be composed with $q \leftarrow q_j$ in (4.29).

Furthermore, by considering the following diagram

$$\begin{array}{ccc} \alpha & & \\ \downarrow & \searrow & \\ q & \leftarrow & q_j \\ \downarrow & & \\ p, & & \end{array} \tag{4.30}$$

The observation $s: q \to p$ of the entire colony q by p explains, by the composition $s \circ \varepsilon_j \circ \delta^{q_j}$, why each individual q_j behaves as it does. In other words, information morphisms from α are compatible with (or relevant to) δ^q and $\{\delta^{q_j}\}$, i.e., the commutativity of the triangular part in diagram (4.30). In this case, the behavior observation of a colony of ants as a global object is interpreted as the composition $s \circ \delta^q$ of morphisms.

We will let $\{\delta^j\}_j = \{\delta^{q_j}\}_j$. Then we glue objects $\{\delta^j\}_j$ in S˜ by an isomorphism between δ^j and δ^k over $q_j \times_q q_k$ in S^.

Note that the descending condition $\delta^j = \varepsilon_j^{*}\delta^q$ over q_j, where ε_j: $q_j \to q$, is satisfied since we already know the global phenomenon δ^q of a colony.

Remark Notice that Type II formulation can be applied to any emerging entity from individuals including our human society.

Type III

As the archetypical example of Type III, we can consider a brain. A brain consists of cells. A brain also consists of lobes which are already self-organized global objects.

We will let b be an ur-presheaf corresponding to a brain. Consider a coproduct

$$\varepsilon_k : b_k \to b = \coprod b_k.$$

of $\{b_k\}_i$, indexed by a finite set. For the commutative diagram

$$\begin{array}{ccc} \alpha & & \\ \delta^b\downarrow & \searrow \delta^k(V_k) & \\ b(\text{-}) & \leftarrow & b_k(V_k) \\ & & \downarrow s(V_k) \\ & & p(V_k), \end{array} \tag{4.31}$$

When b_k is observed by p over $V_k \in Ob(\mathrm{S})$, we have

$$s(V_k) : b_k(V_k) \to p(V_k). \tag{4.32}$$

Then the composition $s(V_k) \circ \delta^k(V_k)$ corresponds to the brain image of the activity of the portion b_k of the brain b over V_k. Since the global b itself need not be defined over V_k, the co-projection morphism ε_k may not be defined over V_k. Note also that the composition $\varepsilon_k(V_k) \circ \delta^b(V_k)$ does not exist. That is, one cannot obtain a global thought δ^b over V_k. As a consequence, observation $s(V_k) \circ \delta^k(V_k)$ of b_k by p over V_k does not provide a global thought δ^b. Namely, we need the notion of descent for data $\{\delta^k\}_k$. For 'gluing' local data $\{\delta^k\}_k$, see 2.5 in Chap. 2.

Then global consciousness or thought δ^b of a conscious entity b induces decomposed morphisms $\{\delta_{kji...l}\}_{kji...l}$ at local levels all the way from various lobes, neurons, atoms, ..., to elementary particles. For all indices $k, j, i,..., l$, morphism δ^b restricts well in the following sense. We have $\delta_{kji...l} = \varepsilon_{kji...l}{}^* \circ \delta^b$ in the diagram:

$$\begin{array}{ccc} \alpha & & \\ \delta^b\downarrow & \searrow \delta^{kji...l} & \\ b(\text{-}) & \leftarrow & b_{kji...l}(V_{kji...l}), \end{array} \tag{4.33}$$

where $\varepsilon_{kji...l}$: $b_{kji...l} \to b = \coprod b_{kji...l}$.

4.4 Synopsis

Beginning with Penrose's Triangle of Three Realms, we cross the line between the physical realm and the mental realm. Our focus in this chapter is on the delicate issues of mind and matter. Avoiding ambiguity, we define explicitly what consciousness is using temporal topos language. Then we are faced with the avalanche of enigmatic issues to be treated in a unified way such as measurement phenomena, free will, consciousness (in 4.1), perception, and the animate in relation to the inanimate (in 4.2). The difference between a canonical morphism, i.e., functorially induced, and a non-canonical morphism, plays crucial roles for the definitions of being animate and being inanimate. In the application Sect. 4.3, we divide the temporal topos applications into three types I, II, and III.

Chapter 5
Philosophical Roles of Temporal Topos

Contents

G. C. Kato, *Temporal Topos Methods for the Philosophy of Natural Sciences*,
https://doi.org/10.1007/978-981-96-2420-1_5

Our interpretation of the following quotations

Music is my life, and my life is music. Anyone who does not understand this is not worthy of God.

Wolfgang Amadeus Mozart

The eye with which I see God is the same eye with which God sees me. My eye and God's eye is one eye, and one seeing, and one knowing, and one love.

Meister Eckhart in The Essential Sermons, Paulist Press, 1981

reflects the role of an initial object α of temporal topos S^. In this chapter, we will explain why the properties of an initial object α of a temporal topos S^ play the role expressed in the comments by Wolfgang Amadeus Mozart and Meister Eckart.

Since all initial objects are isomorphic to each other, they are *fundamentally the same*; however, they are distinguishable. There are precise mathematical concepts to distinguish the difference between 'being exactly the same' and 'being essentially the same.' They correspond to the mathematical notions of an equality and an isomorphism, respectively. We explicitly describe the difference between them as follows. Two objects A and B being equal, i.e., $A = B$, is referred to as 'being indistinguishably the same,' and on the other hand, two objects A and B being isomorphic, written as $A \approx B$, is referred to as 'being essentially the same,' but A and B are distinguishable, i.e., $A \neq B$. Namely, the notion of an isomorphism enables us to differentiate *being essentially the same* from *being exactly the same.*

Note See Appendix I for the definition of an isomorphism. For a proof asserting that there is an isomorphism among all the initial objects, see (1) of Remark following the definition of an initial object.

With this notion of an isomorphism, we can assign traditionally established names through the histories of the West and the East to an initial object α of S^. That is, in addition to calling α the source of perception, an initial object α can be associated with Brahman in Hinduism, and with Ultimate Reality in Taoism in Eastern philosophy. If another initial object α^* is associated with the One in a Neoplatonism's sense, with Hegel's Unifying Absolute in Western philosophy, then α and α^* become isomorphic. When there is an isomorphism between α and α^*, we write $\alpha \approx \alpha^*$.

Let $[\alpha]$ be an equivalence class consisting of all the initial objects isomorphic to an initial object α of temporal topos S^. Then, $[\alpha]$ is unique as a class; however, if α_j and α_k are arbitrary objects, called *representatives*, in class $[\alpha]$, where j and k are positive integers, $j, k \in \mathbf{Z}^+$, then α_k and α_j become isomorphic initial objects of S^. Consequently, we can say that, as an equivalent class consisting of all the isomorphic objects, there is only one class $[\alpha]$. If we regard class $[\alpha]$ as *one*, or if we regard *representatives* of class $[\alpha]$ as *many*, we can assign each case as monism or pantheism, respectively.

An existence of an initial object α *of category* S^ *represents the existence of the omnipresent and omniscient being, which is a consequence of* α *being a constant ur-presheaf (i.e., independent of the ur-site* S, *generalized-time category) and the unique morphism* $\delta^{(-)}$ *from* α *to any object* '-' *of* S^. *Note that* $\delta^{(-)}$ *does not exist until an object in S^ is specified, i.e., replacing* '-' in $\delta^{(-)}$ *by an object of* S^.

In what will follow, *metaphysics*, which means beyond the physical realm involving a terminal object ω, is referred to as a study of *the ultimate reality* which means we need to include the cognitive realm. *Ontology*, as a subcategory of metaphysics, is for the study of existence, which is relevant to measurement or observation (hence epistemology) in the quantum physical sense in Chap. 3. That is, any formulation for physical and cognitive realms must be simultaneously consistent. We express the mutual consistency for both physical and mental realms as the commutativity of diagrams. We are aware that we live simultaneously in the cognitive and physical realms. A philosophical issue enters the concept of observation (since usually there is a physical distance between the observed and the observer) where a self-conscious being and the thoughts are together, that is, self-observation is an auto-observation morphism.

Since metaphysics is the study covering both cognitive and physical realms, it can be said that we study *the ultimate reality*.

The great thinkers of the East and the West are not only fundamentally correct but also far-reaching. The only thing missing among these great accomplishments is a mathematical language to express their thoughts with rigor.

As profound human beings for such accumulated works from the ancient through the modern periods, we mention only such well-known thinkers as Pythagoras, Plato, Laozi, Siddhartha Gautama (The Buddha), Meister Eckhart, Descartes, Spinoza, Leibniz, Hegel, Schopenhauer, and in the fields of physics and mathematics of the twentieth century Hilbert, Einstein, and Grothendieck.

During the Greek period, Democritus described the Realty, offering the purely physical view of the universe consisting of mechanical fundamental entities like atoms. In Plato's view, however, the cosmos is a living soul, introducing the notion of the Forms. The concept of the Forms in the sense of Plato corresponds to Roger Penrose's (Platonic) Mathematical Mental World, existing outside of the Physical World. The realm of the Forms includes being conscious and being aware. What do we mean by saying one understands anything? Furthermore, one can ask where such a cognitive function, e.g., consciousness, is coming from, or what makes an animate entity possible to be conscious. What is a criterion to distinguish an animate being from an inanimate being? We need to fill the chasm between living things and non-living things if such a chasm exists between them. In our approach, the sheaf theoretic notion of 'gluing local (descent) data,' and the notion of a mnemonic morphism play an important role for the process from the local to the global and from the inanimate to the animate, respectively. The concept of animated or sentient particles which Leibniz called 'monads' as a fundamental unit is relevant to obtaining a global object from locally trivial objects like atoms and monocles.

G. W. F. Hagel (1770–1831) adapted the classical notion of fundamental substances called monads, which are connected to an initial object α and the morphism δ which originates from α. His study laid a foundation for an investigation into the study of consciousness with his notion of the Absolute, corresponding to α, which unifies the interplay between consciousness, corresponding to δ, and experience, corresponding to a mnemonic morphism from the past to the present.

Does mathematics exist independent of human-brain activity? Our approach says 'Yes.' The reason for the affirmative answer is coming from our ontological view in terms of existence of an initial object α in temporal topos S^, which will be fully developed in this chapter. On the other hand, for the physical realm, a terminal object ω of temporal topos S^ plays a role as already seen in Chap. 3.

5.1 Philosophical Interpretation of the Fundamental Diagram

The fundamental commutative diagram (*) is the beginning of what will follow in this chapter.

$$\begin{array}{ccc} \alpha & \rightarrow & \omega \\ \searrow & & \nearrow \\ & q. & \end{array} \qquad (*)$$

In terms of the diagram (*) above, we will provide a unified vision of ontology, metaphysics, and epistemology in traditional philosophies of the East and the West. While various discussions have taken place, they can be seen as different expressions of the isomorphic objects of S^. Instead of being swayed by various "appearances," we should be able to grasp the essence as the equivalence class of isomorphic objects. Through fundamental commutative diagram (*), we may be able to read how the physical aspect and the cognitive aspect of *Reality* are internally related and mutually relevant, as expressed in diagram (*). We would like to begin with fundamentals of philosophies leading to categorical and sheaf theoretic intervention into the explicit formulation of philosophical thought of the West and the East.

5.1.1 What Is Philosophy?

In what sense *things* exist, we must specify what *things* are. Then we must ask how we know those *things* exist. Before formulating ontological and epistemological issues, we need axioms capable of explaining in what sense the word *existence* makes sense.

Philosophy is a field of study that seeks to understand the fundamental nature of Reality, existence, knowledge, values, reason, mind, and language. It addresses essential yet naïve questions about life and the universe striving to provide coherent and also comprehensive answers. Note that philosophy means love of wisdom: *philosophia* in Greek. We have already defined in terms of morphisms what observing phenomena is meant. Through the language of modern mathematics, we will begin our journey for the quest for the ultimate structure of *Reality*.

In what will follow, we will first give a concise background of the history of philosophy viewed from the temporal topos theory.

5.1.2 The Origins of Philosophy

From the perspective of Western history, philosophy is said to have originated with Thales of Miletus, a pre-Socratic Greek thinker who is also considered as the founder of sciences including geometry. Thales is said be the first man who predicted a solar eclipse. Before Thales, the world was predominantly understood through mythology and supernatural explanations. These myths were deeply ingrained in the cultural and religious fabric of ancient civilizations providing explanations for natural phenomena and human experiences.

Thales, however, marked a significant departure from this mythological worldview. He sought the arche, the fundamental principle or substance underlying all things through rational inquiry and observation. Thales proposed that water was the arche, the essential element from which everything emerged. This idea, while seemingly simplistic by today's standards, represented a groundbreaking shift toward rationalism. Thales' emphasis on reason and empirical observation laid the groundwork for what we now recognize as scientific and philosophical thinking. This rational spirit is a key reason why Thales is often considered the first philosopher in the Western tradition.

5.1.3 Western Philosophy and Eastern Philosophy

The distinction between Western rational and Eastern non-rational thoughts and approaches has been a cornerstone in the development of Western philosophy. Rational thought relies on logic, evidence, and systematic reasoning, while non-rational thought is based on mythical phenomena not grounded in empirical evidence. For a long time, Western philosophers viewed rationality as the hallmark of true philosophy. It was the 18th century when the profound insights and systematic philosophies of the East developed in regions such as India, China, and Japan were recognized. Gottfried Wilhelm Leibniz (1646–1716) seems to be one of first great European philosophers who understood Eastern thought and the Eastern way of thinking, e.g., Hinduism, Buddhism, and Taoism. It is known that Father Bouvet, who was also a mathematician, introduced the I Ching to Leibniz.

The West and the East offer alternative and complementary ways of understanding *the ultimate reality* emphasizing aspects like harmony, balance, and the interconnectedness of all things.

5.1.4 The Implications of Context-Dependency and Relativism

The acknowledgment of multiple philosophical traditions highlights the context-dependency of philosophical ideas. Different cultures during different historical periods have developed their unique perspectives and methodologies leading to various understandings of *reality* and *existence.*

Relativism challenges the notion of absolute truth that has been a pursuit of Western philosophy. It suggests that truth and meaning are not fixed but can vary depending on cultural, historical, and personal contexts. It raises questions about the possibility of universal principles.

5.1.5 Addressing the Problems of Relativism: Leibniz's Perspective in "Monadology"

Relativism, with its assertion that truth and meaning are contingent upon individual perspectives, presents significant philosophical challenges. In response to these challenges, the philosopher Gottfried Wilhelm Leibniz offers an intriguing perspective in his seminal work *Monadology*. Leibniz's ideas provide a highly nuanced framework for understanding the multiplicity of perspectives while maintaining a coherent vision of a unified reality. Leibniz states in Section 57 of *The Monadology* (Kshetra Books, 2016, originally published in 1714 when Leibniz was 68 years old):

> *And just as the same city viewed from different sides appears quite different, and is perspectively multiplied, so, the infinite multitude of simple substances, there are given, as it were, so many different worlds, which yet are only the perspectives of a single one, according to the different points of view of each Monad.*

Here, Leibniz employs the metaphor of a city, corresponding to α, viewed from various angles, corresponding to $\{\delta^q\}_{q \in S^\wedge}$, to illustrate the concept of multiple perspectives. Just as a city can look vastly different depending on one's vantage point, so too can the universe be perceived in myriad ways by different entities. Leibniz posits that these entities, which he calls monads, are simple substances that make up the fabric of Reality. Each monad, δ^q for each q in S^, reflects the cosmos uniquely, thus creating a multitude of perceived views of the cosmos.

5.1.6 Monads as Mirrors of the World

Leibniz further elaborates that monads, which correspond to $\{\delta^q\}_{q \in S^\wedge}$ in our approach, function as mirrors reflecting the entire cosmos through the commutativity of diagram (*). Each monad, despite being a simple and indivisible entity, encompasses a unique perspective of the whole cosmos via the commutative diagram (*).

This perspective is shaped by the internal states and properties of the monad allowing it to reflect the cosmos in a distinct manner. Thus, while there are many monads, each offering a unique view, they are all reflecting the same underlying Reality, corresponding to α, through $\{\delta^q\colon \alpha \to q\}$ for various objects $\{q\}$ in S^.

For mathematically inclined readers, as we have formulated earlier, δ^q is the cognitive and sentient ur-state of q. When e is the associated ur-presheaf to an *elementary* particle like an electron, δ^e corresponds to the *elementary monad* of e. Elementary monads are building blocks for constructing a more complex global monad. For a global ur-object q consisting of $\{q_i\}$, i.e., $q = \coprod q_i \leftarrow q_i$, morphism δ^q would be the *global monad* of q, which is, mathematically speaking, the natural transformation (the morphism of functors) between ur-presheaves α and q, i.e., $\delta^q\colon \alpha \to q$.

Leibniz describes the relationship between the monads and the cosmos, corresponding to the fundamental commutative diagram (*). Reflecting a dynamic interplay between the observer q and the observed p, we can express the relationship among an observer q, the observed p, α and ω over generalized-time periods as the following commutative diagram

$$\begin{array}{ccc} \alpha & \rightarrow & \omega \\ \downarrow & & \uparrow \\ p(V) & \leftarrow & p(W) \\ \downarrow & & \downarrow \\ q(V^*) & \leftarrow & q(W^*), \end{array} \tag{§}$$

where $V \to W$ and $V^* \to W^*$ are temporally linear morphisms of t-site S. In the commutative diagram, all the morphisms are canonically induced except the observation morphisms $p(V) \to q(V^*)$ and $p(W) \to p(W^*)$.

This relationship highlights the importance of perspective in shaping our understanding of Reality. Each monad's viewpoint contributes to a rich tapestry of perceptions, yet these perceptions are all grounded in the same *fundamental reality* which is synopsized in the fundamental commutative diagram (*). In essence, monads exist as entities that view the cosmos, functioning as mirrors $\{\delta^q\}, q \in S$, as a morphism from α to q, reflecting the entire cosmos. Leibniz refers to the existing world and the monads that reflect such a cosmos emphasizing the relationship between the observer and the observed. It is easy to imagine that describing this relationship involves precisely the concept of morphisms in S^. Even in this phenomenological world with its diversity, if we look at the diversity from an overarching perspective, we find *the ultimate reality* detached from but associated with the diverse phenomena through the commutative diagram (*). In that realm, there exists a metaphysical diagram structured by relationships constituted by morphisms connecting an individual object q with an initial object α and a terminal object ω. In terms of Leibniz's notion of a monad, we summarize that we regard an initial object α as the source of perception, or the source of a monad, and δ as the pure concept of a monad. Then natural transformation δ^q becomes the monad of q, and the ur-particle state $\delta^q(V)$ of the monad δ^q of q is a reified state of δ^q over a generalized-time period V in t-site S.

Note Since a uniquely determined natural transformation δ does not exist without an initial object α, morphism δ depends upon α. In that sense, δ should be denoted as δ_α. Since δ_α would not exist as a natural transformation without the specification of an object q in S^, $\delta_\alpha{}^{(-)}$ should be called an ur-wave state of δ_α, as q(-) is an ur-wave state of q. Finally, $\delta_\alpha{}^q(V)$ is the ur-particle state of monad $\delta_\alpha{}^q$ of q reified over the generalized-time V. Since for temporal topos theory, we assume the existence of an initial object α, we will not use the notation δ_α. That is, we simply write δ_α as δ.

5.1.7 The Quest for Universality

Despite the rise of relativism, the human spirit continues to seek universality as a common ground that transcends individual perspectives. The desire for universal truths reflects a deeply seated aspect of human nature, the quest to find coherence and connection in the complexity of existence. Even as we recognize the multiplicity of perspectives, the search for underlying principles that can unify our diverse experiences remains a central philosophical endeavor. *For mathematically inclined readers*, the universal view toward the diverse phenomena may be phrased as the quotient notion:

phenomena modulo underlying principle '~'

meaning that two distinguishable phenomena P and P* are fundamentally the same when P and P* share the same underlying principle denoted as '~', i.e., the class [P] of phenomena consisting of all P*s satisfying the relationship P ~ P*. Namely, this classification is a basic approach in scientific methods. Phenomena P and P* are apparently different, but they share something in common. For example, as a phenomenon P, we can consider phenomenon of the earth orbiting around the sun, and as a phenomenon P*, the moon orbiting around the earth. Then both phenomena P and P* share a common underlying principle which is the concept of gravity.

In this sense, the philosophical journey can be likened to looking through a kaleidoscope. The patterns and colors inside may constantly change, offering a myriad of beautiful and unique images. However, there is still only one kaleidoscope (creating various pattern-images) and one person looking through it. Beyond the ever-shifting images lies another reality, one that philosophers continue to explore in their quest to understand the world and our place within it.

Thus, philosophy, in its broadest sense, is a dynamic and evolving field that encompasses a wide range of perspectives and approaches. It invites us to explore the depths of human thought, to question our assumptions, and to seek wisdom and understanding in a complex and ever-changing world.

5.1.8 Constructing a Unified Philosophy

We are to construct a philosophical unity from a comprehensive perspective. We demonstrate that through the fundamental diagram (*), we can provide a unified description of ontology, metaphysics, and epistemology in traditional philosophy. Instead of being swayed by various "phenomena," the cosmos through a unifying fundamental commutative diagram (*) is our approach.

However, from another perspective, it can be said that throughout history, including ancient times, philosophers understood the meaning of the fundamental diagram (*). They explained what they grasped based on the philosophical and religious linguistic conventions accepted in their own times and regions. As will be explained in this chapter, our philosophical explanations of Plato's view are an essential part of diagram (*).

Therefore, we aim to explore the philosophical debates in the history of thought which can be viewed as natural explanations of diagram (*). While ancient people are often misunderstood as mysterious and even superstitious, it will become clear that they were interpreting the meaning of the fundamental diagram (*) using their own language. In other words, they were expressing the truths of the cosmos using the language of their time.

In the following sections, we would like to provide an overview of ontology, metaphysics, and epistemology, and introduce the claims of particularly important philosophers.

5.1.9 Pythagoras's Mysticism and Metaphysics

It was Pythagoras who left an 'important margin' for mysticism within it. He is often regarded as a mathematician, but in reality, he was a mystic who studied mathematics. His approach to mathematics was not merely analytical or practical; it was deeply imbued with a sense of the mystical and the transcendent. His famous view that numbers are the fundamental of *ultimate reality* may be rephrased as

Topos is the fundamental of ultimate reality

in the 21st century. Furthermore, we may reinterpret Pythagoras' motto as both the physical and mental aspects of *ultimate reality*, and his motto can be explained in terms of categories and sheaves.

5.1.10 Pythagoras and the Harmony of the Spheres

Pythagoras also studied music and is known for discovering the musical scale known as the Pythagorean scale. He identified mathematically the relationship between the lengths of the strings of the lyre and the musical scale. This discovery was revolutionary because it demonstrated a profound connection between numerical ratios and sensory experience. Pythagoras is said to have heard the scales of the cosmos. This concept, often referred to as the "harmony of the spheres," postulates that the planets and stars move according to geometry of the universe, producing a symphony of music that, while not audible to the human ear, resonates with the universe's inherent order.

Whether this story is true or not is left to the reader's judgment, but it indicates that Greek philosophers accepted the metaphysical belief that something beyond the five senses lies behind the phenomena of the physical world. Pythagoras's teachings suggest that numerical relationships are not just abstract constructs but are fundamentally embedded in the fabric of Reality. It is certain that music transforms our consciousness and leads us to a sensory world beyond simple perceptual experience. Even if we do not know what Pythagoras heard, he was aware of a metaphysical realm of Reality.

5.1.11 Pythagorean Influence on Western Thought

Pythagoras' influence extended beyond mathematics and music; his ideas permeated the philosophical traditions of ancient Greece and its influences. The Pythagorean belief in the immortality and transmigration of the soul influenced Plato and subsequently the entire Western philosophical tradition. Pythagoras' integration of mysticism and rational inquiry laid the groundwork for the later philosophical synthesis that sought to reconcile the material and spiritual aspects of existence.

As modern reflections on metaphysics, some may dismiss metaphysics as outdated or irrelevant in the face of empirical science. However, contemporary theories in physics, such as string theory, although rooted in discussions connected to physical phenomena, can be said to be modern metaphysics spoken in the language of mathematics. String theory, for instance, proposes that the fundamental constituents of the universe are not point particles but rather one-dimensional 'strings' that vibrate at different frequencies. This theory, much like Pythagorean thought, suggests that the universe has a hidden, deeper structure that is not directly observable but can be inferred through mathematical relationships. The metaphysical realm behind the physical universe is naturally required as a foundation to comprehensively describe the physical realm. Just as Pythagoras saw mathematical relationships as a key to understanding the deeper truths of the cosmos, modern physicists use sophisticated mathematical models to describe phenomena that transcend ordinary sensory experience. See [P. Deligne, D. Kazhdan, P. Etingof, J. W. Morgan, D. S. Freed, D. Morrison, L. C. Jeffery, E. Witten (1991)]. Thus, the quest for understanding the metaphysical aspect of Reality continues, bridging ancient philosophical insights with contemporary scientific exploration.

5.1.12 *The Legacy of Pythagorean Mysticism*

Pythagoras' legacy is a testament to the enduring power of integrating mysticism with rational inquiry. His belief in the interconnectedness of all things through numerical relationships and his vision of a cosmos imbued with harmony and order, continue to inspire thinkers in both the philosophical and scientific realms. By recognizing the value of metaphysical inquiry, we not only honor the contributions of early philosophers like Pythagoras but also open ourselves to a richer, more profound understanding of the universe.

5.1.13 *Plato's Philosophy*

Plato is renowned for reinterpreting the metaphysical cosmos captured by Pythagorean mysticism through a rational spirit and establishing it as a philosophical principle. The theory of Forms is the cornerstone of his philosophical system. This theory is significant because it integrates both metaphysics and epistemology offering a comprehensive framework for understanding reality and knowledge.

5.1.14 *Theory of Ideas-Forms and Anamnesis*

Plato's Theory of Ideas hypothesizes that the material universe we perceive with our senses is merely a shadow or imitation of a higher, immutable Realm of Ideas. These ideas are perfect, eternal, and unchanging, representing the true essence of all things. Plato's epistemology, the study of knowledge, is deeply intertwined with his metaphysical perspective.

In Plato's dialogue 'Meno,' there is a famous scene where an uneducated slave boy solves a basic geometry problem. The fact that the boy, who has no formal education in geometry, can solve the problem is considered evidence that he is recalling something already known to him. This process of recollection is called 'anamnesis' which suggests that learning is essentially the recollection of knowledge that the soul already possesses. One of the examples of anamnesis is the ability of a silkworm to know how to make a cocoon without being taught by the previous generations. Anamnesis links the world of ideas with the phenomenal realm, illustrating how true knowledge transcends sensory experience. In our formulation, anamnesis comes from α through δ, not from a mnemonic morphism. To be more explicit: the unique morphism δ^q from α to an individual $q \in Ob(S^\wedge)$ is responsible for anamnesis. That is, anamnesis does not depend upon a mnemonic morphism from past experiences. We can explicitly give the formulation for anamnesis for the situation as follows. (In a certain sense, a human who is referred to as a genius may have the quality of anamnesis: an ability to deduce a highly non-trivial profound insight and conclusion even from trivial information.)

Let $q \in Ob(S^\wedge)$ and let $V \to W$ be a temporally linear morphism from V to W, where W corresponds to the present time. Then let $j^V{}_W$ be a mnemonic morphism from $q(V)$ to $q(W)$, i.e.,

$$j^V{}_W : q(V) \to q(W).$$

Then we have the following commutative diagram

$$\begin{array}{ccc} & \alpha & \\ \swarrow & \downarrow \delta^q(W) & \\ q(V) \to & q(W). & \end{array} \tag{5.0}$$

The phenomenon of anamnesis is obtained through the morphism $\delta^q(W)$: $\alpha \to q(W)$, not from the mnemonic morphism $j^V{}_W$: $q(V) \to q(W)$. For the definition of a mnemonic morphism, see Sect. 4.1 of Chap. 4.

5.1.15 Allegory of the Cave

Further elucidating a theory expressed in *The Republic*, Plato uses the 'Allegory of the Cave' to argue that the phenomenal realm perceived through our five senses is merely an illusion. In this allegory, prisoners are chained in a cave and can only see the shadows of objects cast on the wall by a fire behind them. These shadows represent the deceptive appearances of the material world. The journey out of the cave into the sunlight symbolizes the philosopher's ascent to the knowledge of the Forms, where the ideas themselves are the true Reality. This allegory underscores the distinction between the deceptive world of appearances and the higher reality of the Forms. For Plato, the cosmos is a living existence with the will corresponding to the left-half of the fundamental diagram (*). The other half of the fundamental diagram (*) corresponds to the mechanical aspect of the Ultimate Reality. This situation will be fully explained in 5.1.24 in terms of the following commutative diagram

$$\begin{array}{ccc} \alpha & \to & \omega \\ \downarrow & \nearrow \sigma_p & \downarrow j^{\omega}{}_q \\ p & \to & q. \end{array} \tag{$\wedge$}$$

5.1.16 Parallels with Eastern Philosophy

Plato's Theory of Ideas contains elements that resonate with Eastern philosophical traditions. In Eastern thought, particularly in Hinduism and Buddhism, the material world is often seen as an illusion (Maya). *Upanishads*, ancient Indian philosophical texts, describe Brahman as the *ultimate reality*, from which all phenomena (Maya) emerge. Similarly, in Chinese philosophy, the concept of Taiji in Chinese (Supreme Ultimate), which precedes the separation into yin and yang, reflects the foundational principle from which the cosmos emanates. Both Western and Eastern philosophies share a profound interest in distinguishing between reality and appearances, i.e., phenomena. Ancient thinkers across cultures recognized that the world perceived through the senses is not the ultimate truth. Instead, they sought to understand the underlying reality that gives rise to the phenomena we experience.

5.1.17 Structural Understanding of Reality and Phenomena

Ancient people were keenly aware of the difference between the deeper reality and physical phenomena (appearance) and were interested in the reality as the source of phenomena. In the context of Western philosophy, this was called the One, the absolute Sprit, the Idea, the Form, and in Eastern philosophy, it was called Brahman, the Supreme Ultimate (Taiji). While the details differ due to the segmentation of the world based on vocabulary, from a broader perspective, it can be said that ancient people structurally understood the relationship between reality and phenomena, which is symbolically expressed as a pair (δ , σ) in our t-topos theoretic approach, where δ is the unique morphism originating from the sentient ur-presheaf α, and σ is the unique morphism to the phenomenal spacetime ur-presheaf ω.

Plato's philosophical system provides a structured approach to understanding the relationship between the material realm and the higher realm of the Forms. This dualistic perspective aligns with the structural understanding of *reality and appearances* found in Eastern philosophies. By recognizing the shared themes and approaches in these diverse traditions, we can gain a deeper appreciation for the universal quest to comprehend the true nature of existence.

5.1.18 Aristotle's Philosophy

Aristotle, despite being a disciple of Plato, rejected his master's Theory of Ideas. This fundamental divergence between teacher and student is famously depicted in Raphael's painting "The School of Athens" in the Vatican. In the painting, Plato points his index finger towards the heavens, symbolizing his belief in the transcendent world of the Forms, while Aristotle gestures towards the earth with his palm facing down, emphasizing his commitment to the empirical world. This contrast encapsulates Aristotle's view that Plato's Theory of Ideas (or Forms) was an unrealistic notion detached from the tangible realm of the reality. Namely, Aristotle criticizes Plato's theory of Forms by saying that form and matter cannot be treated separately, i.e., entangled. This is our vision as well expressed by the fundamental commutative diagram (*).

5.1.19 Aristotle's Concept of Forms and Empiricism

Aristotle did acknowledge the existence of non-material forms. However, unlike Plato, he believed that these forms were not separate from the physical world but were inherent within it. According to Aristotle, Forms exist within individual objects $\{q\}_{q\in S^\wedge}$ that can be perceived through sensory experience. He argued that universal Forms do not exist in a separate realm of ideas but are found within the empirical world, which corresponds to the right half of diagram (*). Therefore, by thoroughly examining individual objects, one can understand their universal and immutable nature.

Aristotle's approach was fundamentally empirical. He emphasized the importance of observation and systematic investigation of the natural world to uncover the underlying principles governing it. This empirical methodology set the foundation for what would later become the scientific method. However, Aristotle also saw the necessity for metaphysics, as certain concepts such as causality, existence, and substance needed to be systematized to interpret and understand the knowledge gained from empirical observation. These metaphysical concepts could not be experienced directly, and thus, Aristotle had to construct his own metaphysical framework to support his empirical investigations.

Aristotle's blend of empiricism and complementary metaphysical views can be observed in modern physics. The spirit of deeply exploring phenomena to discover the fundamental laws behind them aligns with Aristotle's empiricism which focuses on the empirical world. On the other hand, once fundamental laws are discovered, the treatment of these laws through theoretical physics or mathematical physics can be seen as modern metaphysics - a framework for understanding the empirical world.

5.1.20 Aristotle's Comprehensive Vision

Aristotle's comprehensive vision sought to integrate empirical observation with metaphysical reflection aiming to achieve a holistic understanding of the world. This integration is evident in his works where he systematically addresses a wide range of topics from the nature of *reality and knowledge* to the principles of life and human flourishing. Aristotle's ability to combine empirical rigor with profound philosophical insight has made his work enduringly relevant and influential.

In summary, Aristotle's rejection of Plato's Theory of Ideas in favor of an empirical approach, combined with his recognition of the necessity of metaphysics, represents a pivotal moment in the history of philosophy. His philosophy, characterized by understanding the natural world through observation and complemented by a metaphysical framework, continues to influence contemporary fields in philosophy and science.

5.1.21 Neoplatonism's Influence on Religion

Neoplatonism is a philosophical school that developed from the 3rd century CE to the 7th century CE. It became the dominant philosophical ideology, offering a comprehensive understanding of the cosmos and humanity's place within it.

Neoplatonism was not merely a revival of Plato's ideas; it was a grand attempt to absorb and integrate the rich intellectual heritage of Hellenistic philosophy, religion, and literature. Excluding the thorough materialism of Epicureanism and Stoicism, Neoplatonists reflected a millennium of intellectual culture engaging Plato, Aristotle, and Stoic ethics in dialogue with literature, mythology, and religious practice.

The philosophical system of Neoplatonism is based on several fundamental assumptions, the most important of which is the ontological superiority of mental consciousness (nous) over physical reality. This premise, shared by many ancient intellectuals, was inherited by Neoplatonism, which developed into an idealistic philosophy.

Another crucial assumption of Neoplatonism is that all reality depends on a single supreme principle referred to as 'the One,' 'the Good,' or 'the First,' which corresponds to an initial object α of temporal topos S^. Neoplatonists believed this supreme principle to be the basis of all reality, which is captured as the commutativity of the fundamental diagram (*). If the assumption that mental consciousness is ontologically superior to physical reality holds, then the first principle must also be a principle of consciousness. In t-topos language, the mental aspect of the consciousness of q is expressed as the universal morphism $\delta^q: \alpha \to q$ followed by the physical reality $\sigma_q: q \to \omega$, whose composition $\sigma_q \circ \delta^q$ is $\zeta: \alpha \to \omega$.

Under these assumptions, Neoplatonists tackled the problem of how the universe emerges from a single principle of consciousness. They believed the generation of the universe to be an eternal process with Reality emerging in stages from the One principle. This process was thought to have a dual nature of internal and external activities where internal activity generates external activity leading to a chain of further external activities.

Note An initial object α and a terminal object ω are dual concepts in the following sense. In the dual category $S^{\wedge opp}$ of S^, an initial object α of S^ becomes a terminal object of $S^{\wedge opp}$, and a terminal object ω of S^ becomes an initial object of the dual $S^{\wedge opp}$.

In Neoplatonism, the external activity of the first principle is pure consciousness (or *nous*, meaning *the mind*), the highest form of being. The internal activity of this consciousness is self-understanding, and through this recognition of *cause*, the overall structure of the universe becomes apparent. From this activity of consciousness, the soul (psychê) arises, functioning as a mediator that enables the emergence of the material world. Namely, consciousness δ^q is the mediator between α and q. Then the emergence of the material world is expressed as the morphism $\sigma_q: q \to \omega$.

Regarding the relationship between the soul and nature, i.e., (α, δ) and (ω, σ), respectively, Neoplatonism speculates that the soul shapes matter. The individual soul receives information through consciousness and reflects it onto the material world. Through this process, the material world becomes an external manifestation of the activity of consciousness. We can regard morphism $\zeta: \alpha \to \omega$ from the One to the phenomenal realm ω to be the reflection of mind manifested upon the external physical realm ω.

Neoplatonism viewed matter as a byproduct of the dynamic emission of the One, with matter existing as a result of the soul's activity. Consequently, matter itself was considered formless and non-substantial with the origin of evil arising from an excessive focus on matter. Note that we have no device to express 'evilness.'

In ethics, Neoplatonism taught that the human soul should aim to return to divine consciousness, i.e., α. Traditional virtues were important, but the ultimate goal was to deepen the relationship with the divine α via δ.

Neoplatonism significantly influenced later Christian, Islamic, and Jewish theologies and contributed to the revival of Western thought during the Renaissance. Its influence persists to the present day playing a vital role in idealistic philosophy.

5.1.22 Kant's Synthesis of Rationalism and Empiricism

Immanuel Kant (1724–1804), whose work was the origin of critical philosophy, critiqued with rigor what had been done in philosophy before him in a coherent and comprehensive way. His aim was to give a solid foundation for unifying cognitive-subjective and physical-objective realms.

Kant's philosophy, particularly as articulated in his seminal work *Critique of Pure Reason*, sought to mediate the long-standing debate between rationalism and empiricism. Rationalism, as exemplified by philosophers like Descartes and Leibniz, asserts that knowledge is primarily acquired through reason and innate ideas independent of sensory experience. Empiricism, championed by thinkers such as Locke and Hume, contends that knowledge is derived from sensory experience and that the mind at birth is a *tabula rasa*.

Kant aimed to bridge this gap by integrating key elements of both Platonic rationalism and Aristotelian empiricism into a cohesive philosophical system. He argued that both rationalist and empiricist perspectives are essential for acquiring knowledge. Kant proposed that while knowledge is indeed grounded in experience, the structural framework that makes experience possible exists independently of that experience. This framework is imposed by the mind itself which actively shapes our sensory inputs through innate structures. These innate structures, which Kant called categories of understanding, include concepts such as causality, space, and time. These categories are not derived from experience but are preconditions for any experience to be intelligible.

5.1.23 The Concept of Thing-in-Itself, Noumenon

A central concept in Kant's philosophy is '*the thing-in-itself*' (Ding an sich), or in the plural form Noumena or 'things-in-themselves,' which refers to the reality that exists independently of our sensory experience. For Kant, physical objects, as dealt with in the sciences, are not the true reality but rather phenomena, i.e., images or appearances shaped by our perceptual faculties. According to Kant, physics is a system that describes these phenomena as they emerge from the underlying Reality. '*The thing-in-itself*' remains inaccessible to direct perception and cognition; we can only know it indirectly through the phenomena it produces. We are naturally led to associate Kant's *noumenon* with Plato's realm of the Forms, which is captured as α in our approach.

Kant presupposed that while we can have extensive knowledge about the appearances of things (phenomena in the right-half realm of the fundamental diagram (*)), the true essence of Reality (*the thing-in-itself*) remains beyond our reach. This distinction is crucial for understanding the limits of human cognition. Phenomena are the objects of our experience mediated by our senses and the categories of our understanding, whereas *the thing-in-itself* is the reality that exists independently of our perceptual and cognitive faculties. We can never know *the thing-in-itself* directly, but it is the source of the phenomena we experience. That is, it is the role played by both an initial object α and a terminal object ω, whose interaction is expressed in the commutativity of the fundamental diagram (*).

In the following remark, we will give a temporal theoretic formulation for Kant's impossibility of knowing *the thing-in-itself* directly.

Remark For mathematically inclined and mathematically uninclined readers

Mathematics expresses its own thought in the following sense. As we noted, an observation morphism is a non-canonical morphism. Therefore, to be able to directly observe α, there would be a non-canonical morphism from α to an observer q, which contradicts α being an initial object. Recall that δ is the unique morphism from α to an object of S^. Consequently, one cannot directly observe α, *the thing-in-itself*. On the other hand, however, spacetime ur-presheaf ω can be observed without contradicting that ω is a terminal object of temporal topos. This is because the unique morphism σ_q is from q to ω while a non-canonical observation morphism of ω is from ω to q.

Furthermore, we will provide a comment *for mathematically inclined readers* on the impossibility of a direct observation of α. Consider the following diagram

$$\begin{array}{ll} \alpha & \\ \downarrow & \searrow \\ \alpha & \rightarrow q, \end{array} \qquad \text{(Kant)}$$

where the vertical morphism is an identity morphism id_α on α, the horizontal morphism is an observation morphism $j^\alpha{}_q$ of α, and the tilted morphism is the uniquely determined δ^q from α to q. Then we have $j^\alpha{}_q = \delta^q \circ id_\alpha = \delta^q$. Consequently, an observation morphism $j^\alpha{}_q$ is nothing but the canonically induced unique morphism δ^q. Therefore, q cannot non-canonically observe α directly.

Might we admit that this kind of harmony is expected between Kantian *the thing-in-itself* concept and temporal topos theory?

5.1.24 Kantian Framework and Representation

Kant's philosophical framework asserts that Reality exists independently of human perception, but what we can recognize are merely phenomena. We cannot directly perceive Reality itself as we have given a t-topos theoretic proof for the assertion in the above Remark. To represent Kant's argument in terms of our fundamental diagram (*), we can outline the relationships between *the thing-in-itself* α, the perceiving subject q, and the phenomenon as follows.

First, *the thing-in-itself* corresponds to an initial object α of temporal topos S^. Namely, as shown in the Remark, *the thing-in-itself* exists independently of our perception and is beyond our experience and cognition, and furthermore, an initial object α of temporal topos represents the true reality that is the source of all phenomena exposed in the commutative diagram (*).

A perceiving subject corresponds to an object q in S^. The perceiving subject q receives information from *the thing-in-itself* α through the unique morphism δ^q: α → q, indicating the process by which the perceiving subject q receives information from *the thing-in-itself* α.

The realm of phenomena corresponds to a terminal object ω of S^, or rather the right-half of diagram (*). The phenomenon appears to the perceiving subject as an observable phenomenon in spacetime ur-presheaf ω. This relationship can be represented as σ_q: $q \rightarrow \omega$ in the fundamental commutative diagram

$$\begin{array}{ccc} \alpha & \rightarrow & \omega \\ & \searrow \; \nearrow & \\ & q, & \end{array} \qquad (*)$$

Showing how *the thing-in-itself* α appears as a phenomenon in terms of the morphism σ_q: $q \rightarrow \omega$.

According to this organization, in Kant's philosophy, *the thing-in-itself* α is the ultimate source of information. The perceiving subject q receives this information through δ^q, which then appears as the phenomenon in ω via σ_q: $q \rightarrow \omega$

that we perceive. The perceiving subject, through the innate structures of the mind, actively interprets and organizes the sensory input received from *the thing-in-itself* α, resulting in the phenomena we experience. We will analyze the interplay among the observed, the observer, α and ω more generally later in 5.2, diagram

$$\begin{array}{ccc} \alpha & \rightarrow & \omega \\ \downarrow & \nearrow_{\sigma_p} & \downarrow j^{\omega}{}_q \\ p & \rightarrow & q, \end{array} \qquad (\wedge)$$

as an extended version of the fundamental diagram (*), which may more explicitly clarify the above argument in a comprehensive way. Note that an ur-presheaf q is the perceiving subject of the observed p via a non-canonical observation (or measurement) morphism $j^{\omega}{}_q$ from ω to perceiver q about observed p. Then the information morphism $j^p{}_q$ from p to q is expressed as $\sigma_p \circ j^{\omega}{}_q$. Note that $j^p{}_q = \sigma_p \circ j^{\omega}{}_q$ is the commutativity statement of the right-lower triangle in (^) above.

5.1.25 Physics and Philosophy

Kant's philosophy has had a profound influence not only on subsequent philosophical discourse but also on the development of physics. In particular, notable physicists such as Albert Einstein and Niels Bohr grappled with the interpretative problems of quantum mechanics through a framework that was heavily influenced by Kantian philosophy. Before delving into the philosophical issues inherent in physics, it is important to recognize that philosophy is fundamentally a discipline that explores foundational questions. Central to philosophy's approach to these profound questions are the fields of ontology, metaphysics, and epistemology.

Ontology and metaphysics are sometimes conflated, yet it is crucial to understand the nuanced distinction between them. Ontology seeks to understand the nature of being, addressing questions about what it means 'to exist' and what entities actually exist in the world around us. Ontology examines the categories of being and their relationships, striving to map out the fundamental structure of Reality.

Metaphysics, while encompassing ontological inquiries, extends to address broader and more abstract concepts. It deals with the nature of Reality at a level that transcends the physical world. Metaphysics investigates entities and principles that go beyond physical existence, including phenomena such as time, space, causality, consciousness, and free will. It seeks to understand the underlying nature of Reality and the fundamental principles that govern it.

Epistemology, another critical branch of philosophy, explores the nature, scope, and limits of human knowledge. It seeks to elucidate how we come to understand the cosmos, the processes behind the formation of beliefs, and the justification for these beliefs. Epistemology addresses questions about the nature of knowledge itself, the sources of knowledge, and the criteria for what constitutes justified belief.

Metaphysics, due to its abstract and speculative nature, has often been criticized as being akin to fiction. Indeed, the history of metaphysics includes numerous debates that have, at times, been driven by imaginative excess rather than empirical evidence. However, genuine metaphysical inquiry lies behind the conceptual foundation of physics. The kind of metaphysics that can be fruitfully connected with physics is a rigorous investigation into the fundamental nature of Reality. In fact, Einstein articulated this connection between metaphysics and physics in a letter to Schrödinger, dated June 19, 1935:

> *The real difficulty lies in the fact that physics is a kind of metaphysics; physics describes 'reality.' But we do not know what 'reality' is; we know it only by means of the physical description.*

> Howard, D., *Holism, separability and the metaphysical implications of the Bell experiments.* In J. Cushing & E. McMullin (Eds.), Philosophical consequences of quantum theory: Reflections on Bell's theorem (pp. 224–53). Notre Dame: University of Notre Dame Press, 1989.

This statement highlights the intertwined relationship between metaphysics and physics suggesting that our understanding of reality is deeply contingent upon our physical descriptions of it. The metaphysics we truly need is literally meta-physics, a 'grounded metaphysics' that can connect with and inform our physical theories.

In light of this perspective, it is valuable to consider the interpretation of quantum mechanics through the lens of quantum information theory. Quantum mechanics, with its counterintuitive phenomena and probabilistic nature, presents profound philosophical challenges. Quantum information theory provides a framework for understanding these phenomena by focusing on the informational content of quantum systems and the principles governing their behavior. By examining quantum mechanics through this theoretical lens, we can gain deeper insights into the nature of Reality and the foundational principles that underpin our physical descriptions of the world. We recommend [Isham-Butterfield (1999)], [Isham-Butterfield (2000)].

5.1.26 Wheeler's 'It From Bit'

At the Third International Symposium on the Foundations of Quantum Mechanics held in Tokyo in 1989, physicist John Wheeler delivered a groundbreaking talk titled 'Information, Physics, Quantum: The Search for Links.' In this influential presentation, he articulated his renowned concept encapsulated in the phrase 'It from Bit.'

Wheeler's idea is summarized in his own words:

> *I, like other searchers, attempt formulation after formulation of the central issues and here present a wider overview, taking for a working hypothesis the most effective one that has survived this winnowing: It from Bit. Otherwise put, every it — every particle, every field*

of force, even the spacetime *continuum itself — derives its function, its meaning, its* very existence*entirely — even if in some contexts indirectly — from the apparatus-elicited answers to yes or no questions, binary choices, bits.*

It from Bit symbolizes the idea that every item of the physical world has at bottom — at a very deep bottom, in most instances — an immaterial source and explanation; that what we call reality arises in the last analysis from the posing of yes-no questions and the registering of equipment-evoked responses; in short, that all things physical are information-theoretic in origin and this is a participatory universe.

J. A. Wheeler. *Information, Physics, Quantum: The Search for Links.* In Proc. 3rd Int. Symp. Foundations of Quantum Mechanics, Tokyo 1989, S. Kobayashi, H. Ezawa, Y. Murayama, S. Nomura, eds., Phys. Soc. Japan, Tokyo, (1990), pp. 354–368.

Wheeler's proposition represents a significant reinterpretation within quantum information theory, drawing inspiration from the quantum interpretation associated with his mentor, Niels Bohr, famously known as the *Copenhagen Interpretation.* This approach emphasizes the role of the observer in the formulation of physical reality.

The principle 'to be is to be perceived' (*esse est percipi*) of Bishop George Berkeley (1685-1753) aligns closely with Wheeler's 'It from Bit.'

In this light, Wheeler's statement is revolutionary as an epistemological perspective in physics, yet it leaves unresolved questions concerning ontology, specifically the origin of "Bit," or the source of information. This ontological question remains a profound challenge.

Both Bohr's interpretation of quantum mechanics and Wheeler's '*It from Bit*' avoid direct references to *the ultimate reality*. Since Immanuel Kant's epistemology, there has been a philosophical tradition asserting that '*the thing-in-itself*' is beyond our direct knowledge; what we can comprehend are phenomena. According to Kantian thought, direct knowledge of Reality is impossible.

Despite this, Kant himself had to offer the concept of '*the thing-in-itself*' as the underlying cause of phenomena. This necessity underscores the inescapability of ontology and metaphysics for us. Consequently, the philosophical challenges inherent in Kant's epistemology also permeate Wheeler's 'It from Bit.'

We believe these philosophical issues can be addressed through mathematical descriptions *linked* to physical theory, a concept we refer to as mathematical metaphysics. This approach proposes a bridge between metaphysical concepts and physical phenomena through coherent mathematical frameworks.

Metaphysics, however, has been criticized by Bertrand Russell and the logical positivists who followed him. Yet, rooted in Aristotle's writings, metaphysics literally means 'beyond physics.' Metaphysics is intended to serve as a foundational discipline that forms a foundation for physics, aiming to describe the phenomena observed in the physical realm. True metaphysics should thus connect to physics, providing a unified structural understanding of the laws and principles underlying physical phenomena.

In this context, Wheeler's 'It from Bit' implies the following:

(1) Information Input: There is an input of information $\delta^q \colon \alpha \to q$ from the information source α to the receiver q.

(2) Information Processing: The receiver q processes the information and outputs it $\sigma_q: q \to \omega$. This represents the process by which the receiver q interprets and processes the received information.
(3) Output Presentation: The processed information is presented as the output result in ω. This indicates how the processed information appears as the final output in the physical realm in ω.

These steps emphasizing the fundamental idea that physical reality and phenomena are fundamentally informational in nature and arise from the interaction and processing of bits of information within a participatory realm. This is the depiction that our fundamental diagram (*) captures.

5.2 Where Mathematics and Philosophy Meet

Note that for an object q of S^, our approach to the ontological statement (i.e., ur-entity q to exist) is understood as our fundamental commutative diagram (*) below.

$$\begin{array}{ccc} \alpha & \rightarrow & \omega \\ & \searrow \quad \nearrow & \\ & q. & \end{array} \qquad (*)$$

That is, entity q is to exist *if and only if* q satisfies the commutativity of diagram (*).

Note also that for q to exist, q need not be observed by any object q^* in the following sense. Diagram

$$\begin{array}{ccc} \alpha & \rightarrow & \omega \\ \downarrow & & \downarrow \\ q & \rightarrow & q^* \end{array} \qquad (\text{^})$$

need not be evaluated over any t-site objects. When an observation of q by observer q^* occurs over V and W, respectively, then diagram (^) becomes the commutative diagram as

$$\begin{array}{ccc} \alpha & \rightarrow & \omega \\ \downarrow & & \downarrow \\ q(V) & \rightarrow & q^*(W) \end{array} \qquad (\text{^}V\rightarrow W)$$

where $V \to W$ is a t-linear morphism in S. Note also that $q(V) \to q^*(W)$ in (^V→W) is a non-canonical morphism.

Recall that generally for every object q of a temporal topos S^, there is induced a uniquely determined morphism δ^q from an initial object α to q by the definition of an initial object of a category whether q is reified or not. Notice also that q can be a subatomic particle, e.g., a proton, a neutron, or an electron, and just the same, q may be a macrocosm living being like a human being. An initial object α of t-topos, as a *constant* functor, is an ur-presheaf which is a generalized-time independent functor. As a generalized-time independent functor, an initial object α can be interpreted as '*omnipresent*.' Another way to phrase the S-independence is that the existence of an initial object α of S^ is free from or beyond generalized time, hence, from the usual time τ which is because ur-functor τ depends upon S as its domain category. Consequently, an initial object α being S-independent implies that α is τ-independent. An initial object α can be expressed as an '*omniscient*' object such that δ potentially induces the unique morphism δ^q from α to a specified object q. Notice that an initial object α may be regarded as the *omnipresent-omniscient*, *universal soul*, or *universal consciousness*, i.e., the traditional concept as a source of mind-soul among Western and Eastern philosophical and theological thoughts.

5.2.1 Remarks on Pair (α, δ)

(1) We have indicated various views and focusses in Western and Eastern philosophies which can be identified in terms of the fundamental commutative diagram (*). Some philosophical approaches emphasize α itself, the class [α], or all the representatives $\{\alpha_i\}$ of [α] as *the omnipresent-omnipotent-omniscient*. Some others pay more attention to the universal morphism δ or $\{\delta^{(-)}\}_{- \in S^\wedge}$ used for the definition of an initial object α. Notice that we earlier wrote $q = q(-)$ for an object q to be an ur-wave state. Hence, without the specification of an object of S^, it is natural for us to call $\delta = \delta^{(-)}$ an ur-wave state of the cognitive natural transformation when an object of S^ is not determined.

(2) The very definition of an initial object α is characterized and expressed in terms of the universal mapping (morphism) property of δ. That means that δ plays a decisive role for characterization of an initial object α. Namely, δ and α are mutually dependent. (See Appendix I for the definition of an initial object of a category and its universal mapping property.) The unique morphism δ^q from an initial object α to an individual object q in S^ may be assigned as e.g., *Atman* in Eastern philosophy or the *Entelechies* in the sense of Leibniz in Western philosophy. We can consider a pair (J, δ^J) as the awakened being or, e.g., a living being having reached *Satori* or *ultimate enlightenment*, which may be phrased as δ^J becoming isomorphic over generalized-time objects of a t-site S. For q, a purpose of the effort or training in q's philosophical or theological discipline, e.g., in Zen or Hinduism, is to make morphism $\delta^q(V)$ closer to an isomorphism from α to $q(V)$.

(3) As we indicated in (1) above, for the categorization of thoughts in terms of t-topos with α and ω, philosophical schools and theological disciplines mostly emphasize α itself, or the class [α] of isomorphic initial objects. Either α or its class [α] has been identified as the Source, the Arche, the Subline and the

Beautiful, the Origin, God, the Ultimate One. Notice that some other schools emphasize

(i) the pair (α, δ),
(ii) $\{\delta^q\}$ for all q in S^,
(iii) an ur-wave state $\delta^{(-)}$.

For materialists, the emphasis may be only on

(a) $\{\sigma_q\}$ for all q,
(b) a pair $(\{\sigma_q\}, \omega)$, or
(c) the right-half of the commutative diagram (*).

To capture *the Reality*, our choice would be the commutative diagram (*) itself. To the best of my knowledge, there do not appear to be any established schools of thought either in the West or in the East paying primary attention to the universal morphism from α to ω, which we have denoted as $\zeta = \delta^{\omega} = \sigma_{\alpha}: \alpha \to \omega$.

For example, the notion of the Trinity in Christianity may correspond to the fundamental commutative diagram (*) in the following sense. Since α is a constant functor with respect to the t-site S, an initial object α is time independent (omnipresent). We can consider that α plays the role for the Divine, and the uniquely determined morphism δ from α (which is induced by an object of t-topos S^) plays the role of the Spirit. Then morphism δ^q would indicate the awareness of q, and furthermore $\delta^q(V)$ would correspond to the conscious (ur-particle) state of q over the generalized-time period V, which can be also understood as q's thought during the generalized-time V. As we noted earlier, the pairs $\{(x, \delta^x)\}_{x \in Ob(S^\wedge)}$ can be regarded as the awakened beings $\{x\}$, whose morphism δ^x is fundamentally an isomorphism from α to x.

For mathematically inclined readers, note that a composition of isomorphisms and the inverse of an isomorphism are isomorphic. Then, the uniqueness argument on an initial object depends upon whether one considers the class of isomorphic objects as one or as several for all isomorphic objects. See the following Remarks and Appendix I.

Note In addition to what we have already mentioned, some may prefer to identify an initial object α of t-topos as *the ultimate, the universal-cosmic consciousness, the absolute spirit*, or *the cosmic soul*. Our emphasis is on *the importance of the role played by such an α simply as an initial object of a temporal topos* S^, *and that is all.*

Remarks Even though a similar proof is given in Appendix I about initial objects, we include the following for the fluidity of our discussion.

(1) If we indicate the equivalence class $[\alpha]$ of all the isomorphic initial objects of S^, then precisely speaking, *an* Ultimate One is a representative object α in class $[\alpha]$, and *the* Ultimate One corresponds to the equivalence class $[\alpha]$ itself. The concepts of an equivalence class and an isomorphism seem to play

a significant role in philosophy and theology. See Appendix I for the definition of an equivalence class.

(2) Let α be an initial object of t-topos S^ and let q and p be objects of S^, e.g., conscious entities. When p is observed by an observer q, the composition of δ^p with the observation morphism j from p to the observer q becomes a morphism from α to q via p, where δ^q is a unique morphism from α to q as noted earlier. Then, we require that the diagram below

$$\begin{array}{ccc} & \alpha & \\ \swarrow & & \searrow \\ p & \rightarrow & q \end{array} \tag{5.1}$$

commutes, i.e., $\delta^q = j \circ \delta^p$.

(3) The gesture of *Namaste* (practiced mainly in Asian counties) as a greeting is based on the above commutative diagram (5.1). That is, q recognizes such an α as *the Ultimate One* by observing p by the commutativity of diagram (5.1), i.e., viewing δ^q *omni-existing* α via the composition of two morphisms as $\delta^q = j \circ \delta^p$: $\alpha \rightarrow q$, which should be understood as the image $Im(\delta^p)$ of α in p being observed by q.

In our fundamental commutative diagram for t-topos theory

$$\begin{array}{ccc} \alpha & \rightarrow & \omega \\ & \searrow \quad \nearrow & \\ & q, & \end{array} \tag{*}$$

for any object q in S^, an object q in S^ can be an ur-presheaf associated with an elementary particle or associated with a conscious being. During the process of constructing a global non-trivial object by 'gluing' the local data, we demand the commutativity of the diagram (*) for every gluing stage.

5.2.2 Comment on Time

For the morphism $\pi_\tau \circ \sigma_q$: $q \rightarrow \tau$ in diagram

$$\begin{array}{ccl} q & \xrightarrow{\sigma_q} & \omega = (\kappa, \tau) \\ & \searrow & \downarrow \pi_\tau \\ & & \tau, \end{array} \tag{5.2}$$

where τ is associated with the usual time, if q does not exist, then σ_q does not exist. Consequently, if there are no particles $\{q\}$ in the physical universe, σ_q does not exist. Since it is the same for space ur-presheaf κ, we assert that *time and space do not exist, satisfying the commutativity of diagram* (5.2), *in an empty universe*, i.e., a universe without any particles. Namely, the unique morphisms σ and δ potentially exist and are reified as δ^q only when an object q in S^ is specified. Notice that just as α is characterized (defined) in terms of δ, spacetime ur-presheaf ω is determined by σ, and vice versa in the sense that their existence depends upon each other.

We have let ζ be the uniquely determined morphism defined by $\zeta = \sigma_\alpha = \delta^\omega$ from an initial object α to a terminal object ω, i.e.,

$$\zeta : \alpha \to \omega. \tag{5.3}$$

The diagram (*) should be regarded as the unified diagram for the cognitive realm, i.e., the left-half and the physical realm, i.e., the *ultimate reality*. In our temporal theoretic view, the commutativity of the fundamental diagram (*)

$$\zeta = \sigma_q \circ \delta^q \tag{5.4}$$

might be called an *equation for the ultimate reality for an arbitrary q*, connecting the physical, the right-half of diagram (*), and the mental, the left-half of diagram (*).

Every cognitive aspect δ^b of, e.g., an ur-presheaf b associated with a brain, must be combined with its physical aspect σ_b of b. The commutativity $\zeta = \sigma_b \circ \delta^b$ of diagram (*) should be interpreted as the connection between the thought δ^b and the corresponding brain's physical activities σ_b.

5.2.3 Existence

Let q be any ur-presheaf representing a particle in either microcosm or macrocosm. Recall that such a particle exists if and only if the associated ur-presheaf q satisfies the commutativity of diagram (*).

Hence, our ontology as a philosophy of 'being' implies, as Descartes says, that if *q thinks*, i.e., if δ^q is reified at a certain $V \in Ob(\mathrm{S})$, then q exists, i.e., pair (q, σ_q) exists as a physical entity. A more symbolic expression of Descartes' statement would be '$\delta^q(V)$ implies $\sigma_q(V)$,' Hence, we rephrase it as

Morphism $\delta^q(V)$ exists. Therefore, morphism $\sigma_q(V)$ exists.

For any entity to exist, one cannot separate the cognitive aspect δ^q from the physical existence as σ_q, namely the pair (δ^q, σ_q) of those uniquely determined morphisms needs to be considered together for q to exist. That is, mind and body are two halves of (*); indeed δ^q and σ_q are different morphisms, yet they exist *harmoniously*, expressed as the *commutativity* of diagram (*), which indicates the compatibility of both realms. Hence, one can say that our t-topos theoretic approach is closer to the non-dualistic parallelism of Leibniz.

Our ontological view toward the fundamental diagram (*) for the triple (α, ω, q), satisfying the commutativity $\zeta = \sigma_q \circ \delta^q$, may be close to the monism thought of Spinoza. However, diagram (*) shows both the ultimate α and the physical ω are interconnected in the sense of the commutativity of diagram (*), i.e., as seen in *Pantheism's* view; we see the images of the ultimate α through $\{\delta^q\}$ for all the objects $\{q\}_{q \in Ob(S^\wedge)}$ being imbedded in ω through σ_q.

Diagram (*) should be interpreted as follows. As for the philosophical mind-body issue, the commutativity of diagram (*) implies that the mental state δ^q must agree with the physical state σ_q (the commutativity requirement) and vice versa. The commutativity of (*) with the notion of descent theory is understood as *mutual (or pre-established) harmony* in the sense of Leibniz's *Monadology* (Philosophische Schriften, VI, pp. 607-623; Section 79); Leibniz says,

> *"The souls act according to the laws of final causes through appetences, means, and ends. The bodies act according to the laws of efficient causes or motions. And these two realms, of final and efficient causes, are in mutual harmony."*
>
> *Philosophy of Mathematics and Natural Science*, by H. Weyl, Princeton University Press, 1949, p. 209.

Note Compare the above quoted translation with the following quotation in Kshetra Books, 2006, p.27:

> *Souls act according to the laws of final causes, by appetitions, means and ends; bodies act according to the laws of efficient causes, or the law of motion. And the two kingdoms, that of efficient causes and that of final causes, harmonize with each other.*

That is, the Reality as a whole can be characterized by simple and independent monads $\{\delta^q\}_{q \in Ob(S^\wedge)}$, which are mutually organized through isomorphisms on the 'overlaps' of local descent data.

In this philosophical view, we consequently have the commutativity $\zeta = \sigma_q \circ \delta^q$ of diagram (*) indicating that the existence of any entity represented by q and any state of q must have an agreement from α via δ^q and be consistent with its physical realm via σ_q.

In Weyl's sense, morphisms δ^q and σ_q represent the two realms of final and efficient causes in the above quotation.

Notice that Arthur Schopenhauer's *The World as Will and Idea* (*Die Welt als Wille und Vorstellung*) seems to claim that *will* stands outside of space and time, agreeing with Kant, where our formulation of the *will* is relevant to spacetime ω, a terminal object of S^. That is, *will* needs to be comparable in the sense of our insistence on the commutativity of diagram (*), not only for *will* but also for the physical spacetime aspect.

Note also that the unalloyed concept of a *monad* in Leibniz's sense is also interpreted as δ alone, and then morphism δ^q from α to q becomes the monad *of* q, i.e., the cognitive awareness of particle q.

The macrocosm objects are considered as global objects obtained from local data consisting of monads satisfying descent data which means suitable for gluing objects over the common domains of the data. For Leibniz's view on Reality, all the objects in the cosmos in microcosm and macrocosm consist of monads in the above sense, not accepting the traditional atomism. (It is also interesting to recall that Leibniz might be the first philosopher who was familiar with both Eastern and Western approaches.) It may be more appropriate to call δ^q a *sentient morphism* of q from α (Leibniz might have called it a *monad morphism* from α to q), and σ_q a *physical morphism* of q.

Let us express and reformulate Leibniz's view of a soul of a living being q as follows in terms of temporal topos. Suppose that q consists of $\{q_i\}$, a co-product of $\{q_i\}$, i.e., ε_i: $q_i \to q = \coprod q_i$ as before. For q and for each q_i, we have δ^q: $\alpha \to q$ and $\delta^i \stackrel{\text{def}}{=} \delta^{q_i}$: $\alpha \to q_i$, respectively, so that these monads $\{\delta^i\}$ control the global monad δ^q in a coherent way, i.e., δ^q is determined by $\{\delta^i\}$ and ε_i. This is because we have the equation $\delta^q = \varepsilon_i \circ \delta^i$ from the commutativity of diagram (5.5):

$$\begin{array}{ccc} & \alpha & \\ & \downarrow \quad \searrow & \\ q = \coprod q_i & \leftarrow & q_i. \end{array} \tag{5.5}$$

In our case, we mean that global monad δ^q is determined by local monads $\{\delta^i\}$.

For a more detailed discussion, see Type I in 4.3, Chap. 4.

Rationalist philosopher Baruch Spinoza's (1632–1677) Pantheistic view of the universe of God can be interpreted as α for his God and also his concept of a unique substance (his *monistic* view) together with all the morphisms $\{\delta^q\}_{q \in Ob(S^\wedge)}$ from α to all the objects in S^. Namely, for his view, there exists only one substance (hence monism), i.e., having the following equalities:

the unique substance
= God
= Cosmos.

For our categorical method, the above equalities are inappropriate and not precise enough. We should replace the equalities with an isomorphism $\zeta = \delta^\omega = \sigma_\alpha : \alpha \to \omega$. On the other hand, since for Spinoza the cosmos is God, Spinoza's philosophy can be classified as Pantheism by emphasizing $\{\sigma_q\}_{q \in S^\wedge}$. Notice that there is a 'one to one correspondence,' i.e., $\delta^q \leftrightarrow \sigma_q$, for each q.

Schopenhauer's view of dividing the universe into two parts, i.e., *will* (*Noumena* or *things-in-themselves*) involving α and *physical phenomena* involving ω, is closer to ours, as the fundamental diagram (*) indicates.

If one insists the *non-commutativity* of diagram (*), then such a view is traditionally regarded as *dualism* in the sense that mind and matter are two distinct aspects of Reality. Namely, our t-topos theoretic view is non-dualistic.

Remark If the t-topos theoretic formulation explains the physical world involving ω and the cognitive world involving α, one can ask a simple question whether t-topos S^ is *real* or not in the most naïve sense, i.e., whether t-topos is not only a mathematical model but also whether the t-topos cosmos is real or not.

We can rephrase the question as to whether we have only found a working mathematical model acceptable enough to explain both physical and cognitive phenomena, or there really exists a *topos-cosmos*, which has been introduced earlier as an 'avatar' world of a 'real' world. Namely, we ask whether temporal topos is only a shadow presentation of what we experience physically and mentally, or the physical and mental realms we experience are only the shadows of temporal topos S^.

We consider that

The eye with which I see God is the same eye with which God sees me. My eye and God's eye is one eye, and one sight, and one knowledge, and one love.

reflects the role of α, as an initial object for God, and the role of δ, as the universal morphism from α, the Spirit, and the pair (E, δ^E) as the awakened, where δ^E is an isomorphism of functors over the t-site S.

For example, philosophically speaking, the three entities in the Holy Trinity in Christianity have been considered to be the same. But, in a temporal topos view, they are never exactly the same, not even isomorphic to each together. The triple $\{\alpha, \delta, (E, \delta^E)\}$ of an initial object α, a unique morphism δ, and a pair of an object E with δ^E, respectively, are distinguishable.

5.3 Eastern Thoughts

Let us begin with a Zen type Mozart's remark, which can be understood as corresponding with the profundity in un-written lines of Basho *haikus*.

The music is not in the notes, but in the silence between.

Wolfgang Amadeus Mozart

The shift from being *self-aware* of individual q to being *purely aware* corresponds with the shift from δ^q to $\delta^{(-)} = \delta$. Furthermore, $\delta^{(-)}$ or δ is what Leibniz calls the Entelechies in *Monadology* (in Section 48), i.e., capable of being. As we remarked earlier, $\delta^{(-)}$ has only a potential existence until part '-' of $\delta^{(-)}$ is replaced by an object of S^.

One of the oldest, if not the oldest, documents in the history of India is the *Rig-Veda*. The original meaning of *Veda* is 'to know' or as a noun 'knowledge.' Later, the *Rig-Veda* became a book of singing and worshiping. The central concepts of Eastern thoughts are *Brahman* and *Atman*. However, when regarded as a divine being, it is *Brahma* (梵天).

Note that the notion of *Atman*, as an individual life, is a central concept in the *Upanishads* (or sometimes also called *Vedanta*) established over the period of several hundred years starting around 500 B. C. However, note also that earlier '*asu*' (living spirit, or *prana*) was used before the usage of Atman in the *Rig-Veda*. See [Ranganathananda, Swami (2000)].

The mental state (including such a state as *cognizance of being*) of a conscious entity q over an object V of a t-site S, can be defined as $\delta^q(V)$: $\alpha \to q(V)$, which is a morphism of functors evaluated at an object V in t-site S. The changing mental states $\{\delta^q(V)\}_{V \in Ob(S)}$ of $\{q(V)\}_{V \in Ob(S)}$ of q are parameterized by the objects $\{V\}$ of t-site S. Namely, for different generalized-times V and W, two different cognitive states correspond as

$$\delta^q(V) : \alpha \to q(V) \, and \, \delta^q(W) : \alpha \to q(W),$$

respectively.

For example, if such a mental ur-state is determined by individual neural ur-states of neurons $\{q_k\}$ in an ur-presheaf b associated with a brain, i.e., $b = \coprod q_k$, then the global mental ur-state δ^b is the manifestation of local data $\{\delta^q{}_k\}$. That is, such a global mental ur-state δ^b is obtained from the local objects $\{\delta^q{}_k\}$ by 'gluing' $\{\delta^q{}_k\}$ through isomorphisms over the 'overlaps' of the descent data. See Descent Type I of Chap. 3.

Let us return to the tradition of Eastern philosophy.

Definition An initial object α of temporal topos represents *Brahman* (or any historical equivalent name mentioned earlier).

In the t-topos setting, the existence of a conscious entity q should be grasped as a triple (q, δ^q, σ_q) in the commutative diagram (*) as before, i.e., q is an object of S^ together with the uniquely determined morphisms δ^q from α to q, and σ_q from q to ω.

If there are awakened levels with respect to α, it would depend upon how close δ^q, for $q \in$ Obj(S^), is to being isomorphic as a morphism from α to q. Ideally, as the ultimate goal of *Zen* through meditation (through *Samadhi* 三昧) morphism $\delta^q(V)$ from α to q becomes an isomorphism for $V \in Ob(S)$. Namely, δ^q is an isomorphism of functors, i.e., universally with respect to t-site objects $\{V\}$. For such a universal individual entity q, an isomorphism δ^q is understood as Atman of q. Then, an initial object α and ur-object q become isomorphic, which is called 梵我一如 (pronounced *Bonga-ichinyo* in Japanese) symbolically expressed as $\alpha \approx q$ via isomorphic δ^q, that is, reaching the timelessness of a constant ur-functor α.

If an individual conscious entity q is identified (more precisely being isomorphic to α) over V with the constant functor α, it is sometimes referred to as a *Satori moment* over V. See Type I in Sect. 4.3, Chap. 4. Such an isomorphism between $q(V)$ and α indicates a life of being independent of time, since the constancy of functor α is with respect to even more fundamental time, i.e., objects of a t-site.

A brief comment on Buddhism: one of the main goals of Buddhism is to obtain the state of Nirvana as an ideal state of mind. Namely, it is to delete all the trivial matters caused by the self-attached mind so that one can identify with the existence of just being, neither in the past nor in the future. By not being influenced by either one's own thoughts or by others', and with this detachment of emptiness, the uniquely determined morphism δ^q by object q becomes closer to isomorphic. The undisturbed state of Nirvana is the state where only morphism δ^q from α to q exists without any morphisms targeting q. Then, the unity relationship between an object and a q-induced uniquely determined morphism may be expressed as a mutual relationship $q \leftrightarrow \delta^q$.

The typical Pantheism approaches (e.g., *Shintoism* in Japan, or just about any naïve and basic religious practice in the earlier histories of many regions on earth) emphasize the commutativity of this diagram

$$\begin{array}{ccc} & \alpha & \\ \swarrow & & \searrow \\ p & \rightarrow & q \end{array} \tag{5.6}$$

For observer q and arbitrary object $p \in S^\wedge$, in the sense that an observer q recognizes α by observing p as the composition

$$f \circ \delta^p : \alpha \to q$$

via p as in diagram (5.6). The more clearly q observes p as an avatar of α, the closer the composition $f \circ \delta^p : \alpha \to q$ is to an isomorphism. That is, q recognizes α as existing anywhere by observing $p \in Ob(S^\wedge)$. The Pantheistic approach to *the ultimate reality* is to recognize an initial object α through an object p in $S^\wedge$ in the above sense. In particular, it is easier for q to recognize α when p is the expression of innocent eyes of a little baby, i.e., $f \circ \delta^p : \alpha \to q$.

> *If one feels the need of something grand, something infinite, something that makes one feel aware of God, one need not go far to find it. I think that I see something deeper, more infinite, more eternal than the ocean in the expression of the eyes of a little baby when it wakes in the morning and coos or laughs because it sees the sun shining on its cradle.*
>
> Br. 1990: 282 | CL: 242 From: Vincent van Gogh To: Theo van Gogh Date: The Hague, Sunday, 10 December 1882

5.3.1 Note on Shintoism

We will consider that naïve and basic religious attitudes existed throughout the continents soon after the last ice age about 10,000 years ago, but before 2000 years ago. These were times when organized religions had no political or financial power. In that sense, we use the words 'naïve and basic' in the above to characterize religious and psychological human attitudes of awe toward nature and people.

As a prototype of such a religion (hardly a religion compared with the religions developed much later), we describe Shinto. Note also this type of Pantheism, including Shintoism, differs from Spinoza's Pantheism as shown in what will follow. The naïve and basic shared feelings toward *nature*, *people*, and *gods* (*kami-gami*) seem to be the essence of the way of Shinto. In a way, the Shinto view is like regarding three morphisms in the commutative diagram (*) as essentially isomorphisms, symbolically expressed as $\alpha \approx q \approx \omega$. The empathetic attitude of Shintoism toward the animate and the inanimate seems to come from the feeling of shared backgrounds with *kami-gami*. Consequently, the boundary between the animate and the inanimate becomes nebulous. See [Herbert (1967)] for a comprehensive description of Shintoism. See also [Zeami (originally published in 1424. (2012)] for the concept of *Hana*, i.e., α, in Noh.

Note. One can ask whether an initial object α and q are ur-entangled or not. The answer is yes and no; recall the definition of ur-entanglement of a pair $(\alpha, q)(V) = (\alpha(V), q(V))$ for any V in t-site S. However, since an initial object is a constant (contravariant) ur-functor, the evaluation at any object of t-site is irrelevant. In this sense, an answer to the question of ur-entanglement between an initial object and an arbitrary object of S^ is formally 'yes,' and practically 'no.'

Remarks

(1) As we have mentioned already, *the ultimate reality* is divided into two classes; according to Schopenhauer, one calls the first class *Will* and the second class *Representation.* For an ur-presheaf q to exist (hence, the corresponding physical entity to exist), according to the fundamental commutative diagram (*), q needs to be captured in both the physical and mental realms. We have analyzed the properties of an initial object α shared by various schools of philosophy. Confucianism, Taoism, and Buddhism are essentially the same; for us, these three are isomorphic to each together.

(2) A crucial question is the following. Does an electron have consciousness? We have defined unalloyed consciousness itself as the unique morphism δ originating from α without the specification of an object of S^, in Chap. 4. However, there is a required condition for such a morphism δ^e from α to each object e. For example, let e be an ur-presheaf associated with an electron which is a part of a conscious entity q. Namely, the ur-presheaf e is a component of a decomposition of q, i.e., $q = \coprod e_i$, where for a certain index k, we have $e_k = e$. The answer is 'Yes' because, as indicated earlier, we have in general the equation $\delta^q = \varepsilon_k \circ \delta^k$ from the commutativity of diagram:

$$\begin{array}{ccc} & \alpha & \\ & \downarrow \quad \searrow & \\ q = & \coprod e_k \leftarrow & e_k. \end{array} \tag{5.7}$$

In this case, we can view that these local monads $\{\delta^k\}$ are determined by the global monad δ^q as $\delta^q = \varepsilon_k \circ \delta^k \colon \alpha \to q$, by descending from q. Namely, there is no choice but δ^k to make diagram (5.7) commute in the following sense. The consciousness δ^k of electron ur-presheaf $e = e_k$ is the restriction of global consciousness δ^q to a component e making diagram (5.7) commutative. See also Sect. 4.3 of Chap. 4.

5.3.2 *To Be Isolated or Not to Be Isolated: That Is the Measurement Phenomenon*

Let us continue with the above (2) of Remarks where α is replaced by ω in diagram (5.7). Even though the behavior of an isolated electron is not deterministic because of quantum uncertainty, the electron associated ur-presheaf e as a component of a global ur-presheaf q is *deterministic* in the following sense. The deterministic nature originates from the equation $\sigma_i = \sigma_q \circ \varepsilon_i$ coming from the commutativity of diagram

$$\begin{array}{ccc} & \omega & \\ & \uparrow \; \nwarrow & \\ q = \coprod q_i & \leftarrow & q_i. \end{array} \tag{5.8}$$

Therefore, the ur-state of $q_k = e$ must satisfy the requirement of the restriction of a global object to the *k-th* component $q_k = e$. That is, the restriction of the global classical state of q must coincide with the ur-state of e among the choices in the ur-lightcone in diagram (2.13) in Chap. 2. Consequently, for the non-isolated case as above, the enigma of a measurement phenomenon has a solution in the following sense. *Id Est*, the choice to the future object W at Level - 1 from the present V at Level–0 in diagram (2.13) in Chap. 2 must satisfy the descending condition $\sigma_k = \sigma_q \circ \varepsilon_k$ from the global q to the *k-th* component $q_k = e$. Furthermore, as a special case, we can consider the entire physical universe itself, which would be the case $q = \omega$, regarding a terminal object ω itself as a co-product of all the objects $\{q\}$ of S^, i.e., $\omega = \coprod q \leftarrow q$. Then the measurement phenomenon would be solved by stating that the choice from V in t-site S associated with present time to a future W must be determined for every ur-presheaf, which corresponds to a part (a component of a co-product) of the physical universe. The requirement says that the restriction of the global ω must match the local datum. There is an issue about the case q = ω. Since ω is a constant ur-functor, ω is independent of temporal site S. Therefore, there is no affirmative conclusion we can draw for measurement phenomena since it depends upon whether a constant ur-functor ω has an ur-particle (or an ur-wave) state or not.

Note After the discussion about Yoneda Lemma and the Yoneda Embedding Theorem in Appendix II, we make final remarks for the relationship between the cause-effect principle and measurement phenomenon. *Mathematically inclined readers* are highly encouraged to read Final Remarks on Cause-Effect and Measurement in Appendix II.

A final comment is that we have provided a few comments relevant to the contents of the paragraph in Hermann Weyl's *Philosophy of Mathematics and Natural Sciences*, pp. 300 – 301, which is quoted as follows:

> "Whether or not the view is tenable that the organizing power of life establishes correlations between independent individual atomic processes, there is no doubt that wherever *thought* and the causative agent of *will* emerge, especially in man, that power is increasingly controlled by a purely spiritual world of images (knowledge, ideas). Is it conceivable that immaterial factors having the nature of images, ideas, 'building plans,' also intervene in the evolution of the living world as whole? Some biologists answer in the affirmative and set out

to describe these factors and their workings in closer detail. Henri Bergson has developed his philosophy of *évolution créatrice,* and it is essentially the same doctrine that, amidst all the fireworks of his wit, G. B. Shaw propounds in his play *Back to Methuselah.* The eternal life of pure thought freed from the bondage of matter is, according to this 'metabiological Pentateuch,' the ultimate goal of evolution. Thus speaks Lilith: 'I am Lilith: I brought Life into the whirlpool of force, and compelled my enemy, Matter, to obey a living soul. But in enslaving Life's enemy I made Life's master; for that is the end of all slavery; and now I shall see the slave set free and the enemy reconciled, the whirlpool become all life and no matter.' Scientists in general will be more cautious. As things stand now, the positing of transcendental creative agents possessing the naturel of ideas, whether philosophically dangerous or desirable, is of no help in solving the actual concrete problems of biology. And it is a fact that (unless Lilith tells us) we know nothing of them, at least not in the same manner in which, by interpretative understanding from within, we know of the thoughts and impulses of ourselves and our fellow beings."

Epilogue

Throughout the treatise, one of the main themes has been the relationship between the local trivial and global non-trivial. In music pieces, paintings, literature, and other cultural accomplishments of human beings, each individual note, each individual brush stroke, each individual word, respectively, is trivial. However, the permutations can create the beauty and the profundity of thought. Then each piece of trivial information can be recovered by restricting a global highly non-trivial object to the local.

Full circle:

> *So, the problem is not so much to see what nobody has yet seen, as to think what nobody has yet thought concerning that which everybody sees.*
>
> Arthur Schopenhauer

Using Archimedes' principle as an example, we would like to explain what

> *... thought concerning that which everybody sees*

meants; it is *much* easier to discover Archimedes' principle by observing the overflowing water from Archimedes' bathtub than by noticing the weightlessness when you submerge a plastic bag full of water into a water tank. And it is even harder yet to notice Archimedes' principle by just watching pond water as it quietly exists in front of our eyes. That is, the difficulty is to think what nobody has yet thought by observing what everyone sees. Namely, it is to find a fundamental principle from *the trivial that everybody sees*. A reason for the difficulty may be found in:

> *It may be that they have become too erudite or received too many honors. Such things cause one to lose contact with the essentials.*
>
> Alexandre Grothendieck in *Reaping and Sowing.*
>
> Reflection and testimony on a past as a
>
> mathematician, translation by Childe, version
>
> July 18th, 2022.

G. C. Kato, *Temporal Topos Methods for the Philosophy of Natural Sciences*,
https://doi.org/10.1007/978-981-96-2420-1

Appendix 1
Categories

For humanities inclined readers, we will introduce one of mathematics' abstract fields called the theory of categories. See [Du Sautoy (2003)] and [Frenkel, Edward (2013)] for other topics in mathematics.

For mathematically inclined readers, we will give a deeper insight into a more sophisticated vision for the interplay between categories and sheaves. Even though we have briefly described the notion of a category in Chap. 1, we will provide a concise introduction to category theory. Our presentation of the theory of categories is informal and unorthodox in the sense that this presentation is in a face-to-face lecture format. For a fully developed category text, we suggest readers consult [Schubert (1972)], [Michel (1965)], or [Kashiwara-Schapira (2006)].

The first paper on category theory by S. Eilenberg and S. MacLane was published in 1945 titled *General Theory of Natural Equivalence*, Trans. Amer. Math. Soc. 58, pp. 231–294. Since then, as an organizing and unifying mathematical language, category theory has been applied to several fields in mathematics.

Objects and Morphisms

Definition

A *category* C consists of.

(i) *objects*, which are written as $U, V, W, X, Y, \ldots$

and

(ii) *morphisms*, which are written as $f, g, h, \ldots$.

Then, for a pair of objects X and Y in category C, we have the set of all morphisms from X to Y, which is written as $Hom_{\mathrm{C}}(X, Y)$. That is, we have.

G. C. Kato, *Temporal Topos Methods for the Philosophy of Natural Sciences*,
https://doi.org/10.1007/978-981-96-2420-1

$$Hom_C(X, Y) = \{\text{all the morphisms from object } X \text{ to object } Y\}.$$

If f is a morphism in set Hom_C (X, Y), i.e., $f \in Hom_C$ (X, Y), we write

$$f : X \to Y.$$

Then set Hom_C (X, Y) must satisfy the following conditions (a), (b), (c), and (d).

(a) Equality between two sets Hom_C (X, Y) and Hom_C (U, V), i.e.,

Hom_C $(X, Y) = Hom_C$ (U, V) holds if and only if $X = U$ and $Y = V$.

(b) For morphisms f, and g from W to Y, and Y to V, respectively, i.e.,

$$f{:}W \to Y, \text{ and } g{:}Y \to V$$

in C, the *composition* $g \circ f$ of f and g is defined as a morphism from W to V. Namely, we have

$$g \circ f : W \to V. \tag{1}$$

(c) Furthermore, for f: $W \to Y$, g: $Y \to V$, and h: $V \to U$, the associative law

$$h \circ (g \circ f) = (h \circ g) \circ f : W \to U \tag{2}$$

must hold.

(d) For any object W in C, there is an identity morphism

$$id_w : W \to W \tag{3}$$

satisfying $f \circ id_W = f$: $W \to Y$ for any f: $W \to Y$, and $id_W \circ g = g$: $U \to W$ for any morphism g: $U \to W$.

Definition
Objects X and Y in category C are said to be *isomorphic* if there are two morphisms f: $X \to Y$ and g: $Y \to X$ satisfying $g \circ f = id_X$ and $f \circ g = id_Y$. Then we write $X \approx Y$ or $X \cong Y$ when there is an isomorphism between objects X and Y.

We write the equivalence class of isomorphic objects as $[X]$ to which X belongs. More explicitly, we define $[X] = \{X^{\#} \in Ob(C) | X^{\#}$ is isomorphic to $X\}$. Then any object $X^{\#}$ in the class $[X]$ is said to be a *representative* of the equivalence class $[X]$.

Let us consider two examples of a category.

Example (A)
The category (*Set*) of sets is defined as follows. The set

$$\mathrm{Hom}_{(Set)}(X, Y) \tag{4}$$

consists of all set-theoretic mappings from set X to set Y.

Example (B)
Let (*Vct*) be the category of vector spaces over the set **R** of real numbers (or over the set **C** of complex numbers). Then the set

$$Hom_{Vct}(H, L). \tag{5}$$

consists of all the linear transformations over **R** (or **R**-linear mappings) from vector space H over **R** to vector space L over **R**. Recall that a vector space H is a set satisfying the axioms for a vector space, i.e., H is closed under an operation $+$ in the sense that for any two elements u and v in H, $u + v$ is an element of H, and for any element x in **R** and any element v in H, we have $xv \in H$ satisfying $x(u + v) = xu + xv$ in H.

Therefore, whenever assertin
g that a collection of 'things' forms a category, you need to specify two items:

(i) What is an object of the category, and
(ii) What is a morphism in the category.

Let's begin with (*i*) of the above Example (A). An object of the category (*Set*) of sets is a set consisting of elements. In the second example (B), an object of category (*Vct*) is a vector space over **R**.

Next let's consider (*ii*) for category (*Set*) and category (*Vct*).

A morphism in category (*Set*) is a set-theoretic mapping (or a function). Such a mapping f from a set X to a set Y is written as

$$f{:}X \to Y.$$

The definition of a mapping follows. For each element a in X, there corresponds only one element b in Y, written as

$$f(a) = b^{\circ} \text{in } Y.$$

Then the composition of mappings is the usual composition of mappings defined in terms of elements. That is, for f: $X \to Y$ and g: $Y \to Z$, define the composition $g \circ f$: $X \to Z$ for a in X as $(g \circ f)(a) = g(f(a)) = g(b) \in Z$.

The identity mapping id_X: $X \to X$ is defined as

$$id_x(a) = a$$

for an arbitrary element a in X.

An object of category (Vct) is a vector space over **R**. Let us consider a morphism f from a vector space H to a vector space L. First, f is a set-theoretic mapping from H to L, where H and L are regarded simply as sets, i.e.,

$$f: H \to L.$$

Such a mapping f is called a *linear transformation* of vector spaces over **R**, or simply an **R**-*linear mapping*. Here is the definition of a linear transformation f: we have

$$f(v+u) = f(v) + f(u) \text{ in } L, \tag{6}$$

for any v and u in H. One more condition for f to be a linear transformation over **R** is, for xv in H, where x in **R** and v in H,

$$f(xv) = xf(v) \tag{7}$$

must hold in L. Notice that we get the following equalities from (6) and (7):

$$\begin{aligned} f(xv + yu) &= f(xv) + f(yu) \\ &= xf(v) + yf(u) \end{aligned}$$

Then $Hom_{(Vct)}(H, L)$ becomes the set of all the linear transformations from H to L over **R**.

Definition

Let C be a category. An object α in C is said to be an initial object of C when the following condition is satisfied. For an arbitrary object X in C, there exists a uniquely determined morphism δ^X from α to X. We can express the condition as

$$Hom_c(\alpha, X) = \{\delta^X : \alpha \to X\}. \tag{8}$$

Namely, for any object X in C, set $Hom_C(\alpha, X)$ always consists of exactly one element δ^X.

As the dual notion of an initial object, we can define a terminal object ω of C as an object to satisfy

$$Hom_c(Y, \omega) = \{\sigma_Y : Y \to \omega\} \tag{9}$$

for an arbitrary object Y in C.

Note that this kind of characterization in terms of morphisms of an initial object of a category is called 'a universal mapping (or morphism) property.' In that sense, δ^X may be called a universal morphism defining an initial object α.

Remarks

(1) Should we regard an *initial object* α as '*an*' initial object or '*the*' initial object? Let α^* be another initial object of C. Since α^* is an initial object of C, for an object α of C, there exists a unique morphism δ^α from α^* to object α. On the other hand, since α is an initial object, there exists a unique morphism δ^{α^*} from α to α^*. If we compose those uniquely determined morphisms δ^α and δ^{α^*}, we get a morphism from α^* to α^*. The uniqueness of the morphism for an initial object α^* demands that the composed morphism must be an identity morphism id_{α^*}. And if we begin the same argument beginning with α, the composed morphism must equal id_α as well, proving α is isomorphic to α^*. Consequently, identifying all the isomorphic objects as one object, we should use the article 'the' *initial object* rather than 'an' *initial object*. On the other hand, if we let $[\alpha]$ be the equivalence class of all initial objects, i.e., $[\alpha] = \{\alpha^* | \ \alpha^*$ is equivalent to α in the sense that both α^* and α are isomorphic initial objects of C$\}$, then $[\alpha]$ is the uniquely determined initial class consisting of isomorphic initial objects of C.
(2) Morphism δ^ω: $\alpha \to \omega$ is a uniquely determined morphism from an initial object α to an object ω. Similarly, morphism σ_α: $\alpha \to \omega$ is the uniquely determined morphism from an object α to a terminal object ω. If there is a morphism h: $\omega \to \alpha$, then consider the composition $h \circ \delta^\omega$ as a morphism from α to α. The uniqueness coming from an initial object α implies $h \circ \delta^\omega$ must be id_α. On the other hand, the composition $\sigma_\alpha \circ h$ as a morphism from ω to ω must be id_ω by the uniqueness coming from ω as a terminal object. Namely, such an h must be an isomorphism.

Recall we have used $\zeta = \delta^\omega = \sigma_\alpha$: $\alpha \to \omega$, regarding α as an initial object for δ^ω and regarding ω as a terminal object for σ_α, respectively.

Contravariant Functors

Definition

A *contravariant functor* F from category C to category D is an assignment satisfying:

(i) For any object U in C, $F(U)$ is an object in D.
(ii) Let U and V be two objects of category C and let $f \in Hom_C\ (U, V)$, i.e., f: $U \to V$ in C. Let us define what F does to morphism f: $U \to V$ in C.

By (*i*), F takes all objects in C into objects in D, namely, F takes any object U in C to $F(U)$ in D, and V in C to $F(V)$ in D. That is, a functor is a correspondence between objects of categories. Next, let us explain what the contravariant functor

$$F\colon \mathrm{C} \to \mathrm{D} \tag{10}$$

does for morphisms in category C to morphisms in category D.

Any contravariant functor F must satisfy two properties (F_0), (F_1) and (F_2) as parts of required axioms. First, we have

$$(F_0) \qquad F(V) \to F(U).$$

Let's consider three objects U, V, W in category C, which are connected by two morphisms f and g as follows:

$$f: U \to V \text{ and } g: V \to W \tag{11}$$

If you compose these two morphisms, we get a morphism from U to W, and we write it as

$$g \circ f: U \to W \text{ in } C \tag{12}$$

Then we get directly the morphism $F(g \circ f)$ from $F(W)$ to $F(U)$ in D. Since F is contravariant, we must reverse the order in (12) to:

$$F(g \circ f): F(W) \to F(U) \tag{13}$$

in D. We can split (13) into two steps: first by $F(g)$ from $F(W)$ to $F(V)$ followed by $F(f)$ from $F(V)$ to $F(U)$. That is, we get

$$F(g): F(W) \to F(V)$$

in D, connected by

$$F(f): F(V) \to F(U)$$

in D.

Then we can compose these two morphisms $F(g)$ and $F(f)$, i.e., $F(f) \circ F(g)$ from $F(W)$ to $F(U)$

$$F(f) \circ F(g): F(W) \to F(U) \text{ in category D.} \tag{14}$$

In (13) we obtain morphism $F(g \circ f)$ directly from $F(W)$ to $F(U)$, and in (14) we obtain $F(g) \circ F(f)$ from $F(W)$ to $F(U)$ through $F(V)$.

The first axiom of a contravariant functor is:

$$(F_1) \qquad F(g) \circ F(f) = F(g \circ f). \tag{15}$$

Consider the following in category C associated with (15).

$$\begin{array}{ccc} U & & \\ f\downarrow & \searrow & \\ V \underset{g}{\rightarrow} & W. & \end{array} \tag{16}$$

In diagram (16), we can reach W from U in two ways. That is, the shortcut is from U directly to W, or the 'long' cut is via V. The diagonal shortcut $U \rightarrow W$ is the composition of f and g, i.e., $g \circ f$. Let's take the whole diagram (16) into category D by contravariant functor F. We obtain the following diagram in D, by using the contravariant functor F:

$$\begin{array}{ccc} F(U) & & \\ F(f)\uparrow & \nwarrow F(g \circ f) & \\ F(V) \underset{F(g)}{\leftarrow} & F(W). & \end{array} \tag{17}$$

Notice that the arrows in (16) are reversed in (17) because functor F is a contravariant functor. That is, axiom (F_1) says diagram (17) is commutative, i.e., $F(f) \circ F(g) = F(g \circ f)$.

Next, as the second axiom of a contravariant functor, let us see what a functor F: C $\rightarrow$ D does to the identity morphism, i.e., id_U: $U \rightarrow U$ in category C. For example, if C is the category of sets, then an object U in C is a set. An identity morphism id is an identity map (function) defined by $id_U(x) = x$ for any element x in set U. For a general category C, there may not be an element in an object since an object need not be a set. We need a more general characterization of an identity morphism than the definition of an identity mapping for sets as in (3). That is, the definition of the morphism version is given as follows: the composed morphisms $id \circ f$ and $f \circ id$ are equal to f.

For an identity morphism

$$id_U : U \rightarrow U \tag{18}$$

in C, the contravariant functor F changes the direction from $\rightarrow$ in C to $\leftarrow$ in D. Therefore, we get

$$F(U) \leftarrow F(U) \tag{19}$$

in D. Since $F(U)$ is an object of category D, an identity morphism on $F(U)$, i.e., $id_{F(U)}$: $F(U) \rightarrow F(U)$ exists by the definition of a category.

The second axiom for a functor says that for an identity morphism id_U on U, $F(id_U)$ must be an identity on $F(U)$. Namely, as the image of id_U under F, we must have

$$(F_2) \qquad F(id_U) = id_{F(U)}. \tag{20}$$

Notes

(1) For $f: U \to V$ in C, when F does not change the direction of an arrow $\to$, i.e.,

$$F(f): F(U) \to F(V)$$

in D, then functor F is said to be a *covariant* functor.

(2) For temporal topos theory, we use contravariant functors. A contravariant functor is also called a *presheaf*.

Example As an extremely important example of a contravariant functor, we will examine

$$F = Hom_{\mathrm{C}}(\text{-}, V)$$

from category C to the category (*Set*) of sets. Namely,

$$F = Hom_{\mathrm{C}}(\text{-}, V):\ \mathrm{C} \to (Set).$$

By substituting U into the missing spot '-,' i.e., replacing '-' by U in Hom_{C} (-, V), we have Hom_{C} (U, V), which is the set of all the morphisms from object U to object V in category C. Namely, functor $F = Hom_{\mathrm{C}}$ (-, V) takes an object U in category C to set $F(U) = Hom_{\mathrm{C}}$ (U, V), i.e., an object of (*Set*). I.e.,

$$F = Hom_{\mathrm{C}}(\text{-}, V) \text{ takes } U \text{ to } F(U) = Hom_{\mathrm{C}}(U, V).$$

Notice that functor Hom_{C} (-, V) is indeed contravariant since this functor Hom_{C} (-, V) reverses the direction from

$$f: U \to U^*$$

to

$$F(f) = Hom_{\mathrm{C}}(f, V): Hom_{\mathrm{C}}\big(U^*, V\big) \to Hom_{\mathrm{C}}(U, V), \tag{21}$$

which for $h \in Hom_{\mathrm{C}}$ (U^*, V) assigns a morphism from U to V as

$$F(f)(h) = h \circ f \in Hom_{c}(U, V). \tag{22}$$

It is best to look at the following diagram to see why $F(f) = Hom_{\mathrm{C}}$ (f, V) is defined as in (22).

$$\begin{array}{ccc} & & V \\ & \nearrow & \uparrow_h \\ U & \to & U^*. \end{array}$$

Namely, F is a contravariant functor. If we rewrite (21) using F, we have:

$$F(f){:}\, F(U*) \rightarrow F(U). \tag{21*}$$

In addition to proving the contravariant nature of $F = Hom_{\mathrm{C}}$ (-, V),

we still need to show $F(g \circ f) = F(f) \circ F(g)$ and $F(id) = id_F$, where

$$f{:}\, U \rightarrow U * \text{ and} g : U* \rightarrow U **.$$

Let us prove $F(g \circ f) = F(f) \circ F(g)$ and $F(id) = id_F$.

$$\begin{array}{ccccc} & f & & g & \\ U & \rightarrow & U^* & \rightarrow & U^{**} \\ & & & & \downarrow h \\ & & & & V. \end{array}$$

Let h: $U^{**} \rightarrow V$ be an arbitrary morphism. Then, by the repeated use of (22), we have:

$F(g \circ f)$ (h), by (22) we have.
$= h \circ (g \circ f)$, and by the associative law we have.
$= (h \circ g) \circ f$. Then, by (22) we have.
$= (F(g)(h)) \circ f$. Again by (22) we have.
$= F(f)(\, F(g)\, (h))$. By the definition of composition we have.
$= (F(f) \circ (\, F(g))(h)$.
Consequently, we get $F(g \circ f) = F(f) \circ F(g)$: $F(U^{**}) \rightarrow F(U)$.

Next, for id_A: $U \rightarrow U$, since F is contravariant, $F(id_U)$ reverses the direction formally written as $FU \leftarrow FU$. Then, for any p: $U \rightarrow V$, i.e., p is an element of $F(U) = Hom_{\mathrm{C}}\,(U, V)$, we have $F(id_U)(p) = p \circ id_U = p = id_{FU}(p)$. Consequently, $F(id) = id_F$ holds.

Co-product and Product

Let C be a category and let A_i be objects of C, where i is an element of an index set I, i.e., $i \in I$. A co-product (or coproduct) of $\{A_i\}_{i \in I}$ is denoted as $\coprod A_i$, $i \in$ I and ε_i: $A_i \rightarrow \coprod A_i$. Then the following *universal mapping property* is satisfied: Let A^* be any object of C with morphisms ε_i^*: $A_i \rightarrow A^*$. There is a unique morphism f from $\coprod A_i$ to A^* so that the following diagram commutes:

$$\begin{array}{ccc} \coprod A_i & & \\ \varepsilon_i \uparrow & \searrow f & \\ A_i & \rightarrow & A^*, \end{array}$$

i.e., $\varepsilon_i^* = f \circ \varepsilon_i$. As the dual notion of a co-product, we can define the notion of a *product*, which is not used in this treatise. Readers can consult [Gelfant-Manin (1996)], [Hilton-Stammbach (1971)], or [Kato (2006)].

Appendix 2
Sheaves

Presheaf

We begin with an example.

Example Let C be the category of open intervals, e.g., subsets like $\{x: 2.31 < x < 5.63\}$, $\{x: -7 < x < 3.14\}$, … are open intervals in the set **R** of real numbers. In general, we can write such an open interval in **R** as (a, b), which is the set of all the real numbers x in **R** satisfying $a < x < b$. Namely,

$$(a, b) = \{x: a < x < b\}.$$

In the category C of open intervals, morphisms are inclusion maps only. That is, we define the set $Hom_C\ ((a, b), (a', b'))$ as only consisting of one element, i.e., the inclusion map ι, only if $(a, b) \subset (a', b')$, i.e.,

$$Hom_C\big((a, b), \big(a', b'\big)\big) = \{\mathrm{i} : (a, b) \to \big(a', b'\big)\}.$$

Otherwise, that is, if $(a, b) \not\subset (a', b')$, we define.

$$Hom_C\big((a, b), \big(a', b'\big)\big) \text{ as an empty set } \emptyset.$$

An example of a contravariant functor F from the category C of all open intervals to the category (*Sets*) of sets is as follows.

For an open interval $U = (a, b)$, i.e., an object of C, define $F(U) = F((a, b))$ to be the set of all functions, continuous or not, defined on the open interval $U = (a, b) \subset$ **R**.

G. C. Kato, *Temporal Topos Methods for the Philosophy of Natural Sciences*,
https://doi.org/10.1007/978-981-96-2420-1

As other examples, let.

$$F(U) = F((a, b)) = \{\text{all continuous functions defined on } (a, b)\},$$

or.

$$F(U) = F((a, b)) = \{\text{all differentiable functions defined on } (a, b)\},$$

These function spaces are examples of a contravariant functor

$$F: \mathrm{C} \to (Sets).$$

That is, F takes an object $U = (a, b)$ to an object $F(U)$, which is the set of functions of various types defined on the open interval (a, b) as shown in the above examples.

Next, we will show that functor F from category C of open intervals to the category (*Sets*) of functions is contravariant.

For $(a, b) \subset (a', b')$, i.e., inclusion map

$$\iota: (a, b) \to (a', b'),$$

we have

$$F(\iota): F((a', b')) \to F((a, b)).$$

Notice that $F(\iota)$ is defined by $F(\iota)(f) = f\,|_{(a,\,b)}$, restricting function f over (a', b') to (a, b). In other words, $F(\iota)$ is the restriction of a function f defined over (a', b') to an open interval (a, b) contained in (a', b'). In a way, $F(\iota)$ is an identity mapping, since assignment $F(\iota)$ regards f defined over (a', b') as the same f, but this time f is defined over a smaller interval (a, b) than the original open interval (a', b'). See (*a*) in A.1.1.

Let us return to a general discussion.

Why do we use contravariant functors rather than covariant functors for a temporal topos theory?

One of the reasons to use 'contravariant' is the following.

Let $V \to U$ be a flow of time, i.e., temporally linear, abbreviated as t-linear, for our terminology. It is the temporal site expression for the usual linear time order. For example, let 12:00 o'clock noon correspond to V, and let 13:00 o'clock correspond to U, i.e., U corresponds to one hour later than V. When you observe or measure the state of an ur-presheaf p over generalized time U corresponding to 13:00 o'clock, you have no information about the preceding state of p over V corresponding to 12:00 noon. For example, suppose that something happened to a cat at noon. If you could observe a cat P lying on the ground at 13:00 which is corresponding to the present ur-particle state $p(U)$ of a cat ur-presheaf p, then you cannot know what happened to the cat one hour ago, corresponding to the preceding past state $p(V)$. Namely, the cat might have fallen from the sky, or even the cat could have been knocked down by a rock. However, if you witness a rock hitting a cat's head at 12:00 noon, which is corresponding to the ur-particle state $p(V)$, then you can get some information about the state $p(U)$ of the cat one hour later.

Let us rephrase the above example in terms of morphisms as follows. For

$$f: V \to U, \tag{23}$$

let us see what functor p does to the t-linear morphism $f: V \to U$ in (23).

First of all, we have two objects $p(V)$ and $p(U)$ and the connecting morphism between them:

$$p(f): p(V) \;\cdots\cdots\; p(U). \tag{24}$$

The unspecified direction of the dotted line in (24) is intentional, i.e., the direction of the arrow is either $\leftarrow$ or $\rightarrow$.

When an observer q observes p over a generalized-time period V, information of the state $p(V)$ goes to $q(V)$. Namely, an observation morphism s_V is given by

$$s_V: p(V) \to q(V),$$

i.e., an information flow is from the ur-particle state $p(V)$ to the observer $q(V)$.

In order to get information about the future state $p(U)$, the direction of the arrow in (24) must be the morphism-composable direction, i.e.,

$$p(f): p(U) \to p(V) \tag{25}$$

so that you can continue the flow from the future state $p(U)$ to $q(V)$ via $p(V)$. That is, we can compose these two morphisms of the canonical morphism from $p(U)$ to $p(V)$ followed by the non-canonical observation morphism $p(V)$ to $q(V)$. Consequently, p is a contravariant functor, namely, reversed the direction of $f: V \to U$ to $p(V) \leftarrow p(U)$.

Yoneda Lemma

Next, we will explain the Yoneda Lemma. Let C be a category and let (*Set*) be the category of sets. And let U and V be objects of category C. Let us focus on functor $V^\wedge = Hom_C(\text{-}, V)$. Notice that $V^\wedge$ is a contravariant functor from C to (*Set*) as we have already proved. That is, '^' assigns an object V in C to an object $V^\wedge$ in C^. In other words, assignment ^ is a functor from domain category C to category C^ of contravariant functors as the target category. That is, we have a functor '^':

$$\wedge: C \to C^\wedge$$

In the following, I will discuss more about functor ^, which is said to be *Yoneda's Embedding* functor from C to C^.

Here is the meaning of an embedding of a category into another category. Let F be a contravariant functor from category C to category D. Let $f: U \to V$ be a morphism in C. We already saw that there is induced a morphism $F(f): F(V) \to F(U)$ in D. Namely, when we have the induced set theoretic mapping, we denote the mapping with the letter $\boldsymbol{Y}$ as

$$\boldsymbol{Y}: Hom_{\mathrm{C}}(U, V) \to Hom_{\mathrm{D}}(F(V), F(U)). \tag{26}$$

Note that for $f: U \to V$ in C, the above $\boldsymbol{Y}$ is defined by

$$\boldsymbol{Y}(f) = F(f) : F(V) \to F(U) \in Hom_{\mathrm{D}}(F(V), F(U)).$$

Definition

When $\boldsymbol{Y}$ is an injective mapping from set Hom_{C} (U, V) to set Hom_{D} $(F(V), F(U))$, functor F is said to be *faithful*, and when $\boldsymbol{Y}$ is a surjective mapping, F is said to be *full*. Furthermore, F is said to be *fully faithful* when $\boldsymbol{Y}$ is bijective (i.e., $\boldsymbol{Y}$ is both injective and surjective).Finally, F is called an *embedding* if F is not only a faithful functor but also functor F is injective on objects, i.e., $F(U) = F(V)$ in D implies $U = V$ in C. We sometimes identify U in C with the image $F(U)$ in D when F is an embedding functor.

Definition

A morphism in C^

$$d: F \to G \tag{27}$$

is said to be a *natural transformation* (i.e., a *morphism of functors*, in other words, d is a morphism from object F to object G of category C^) when $G(f) \circ d_V = d_U \circ F(f)$ holds in (30).

Namely, since F and G are functors, for objects U and V of category C, and for $f: U \to V$, we have the following mappings of sets

$$F(f): F(V) \to F(U) \tag{28}$$

and

$$G(f): G(V) \to G(U). \tag{29}$$

Then we get the following diagram by combing (27), (28), and (29).

$$\begin{array}{ccc} F(V) & \to & F(U) \\ \downarrow d_v & & \downarrow d_u \\ G(V) & \to & G(U) \end{array} \tag{30}$$

If the above diagram (30) commutes, $d: F \to G$ in (27) is a natural transformation i.e., a morphism of functors. The commutativity of diagram (30) is given by $G(f) \circ d_V = d_U \circ F(f)$.

Note that functor ^ is a *covariant* functor from C to C^, i.e.,

$$\wedge: C \to C^{\wedge}.$$

Here is a proof of the covariant-ness of functor ^. Let

$$U \to V \tag{31}$$

be a morphism in C. To show ^ is a covariant functor, we need a morphism in C^ in the same direction of (31). Namely, we need to define a natural transformation

$$U^{\wedge} \to V^{\wedge} \tag{32}$$

Let us evaluate those functors at an arbitrary object W in C, that is, let us look at U^(W) and V^(W). By definition,

$$U^{\wedge}(W) = Hom_C(-, U)(W) = Hom_C(W, U),$$

and similarly, V^ (W) = Hom_C (W, V). In the diagram

$$\begin{array}{ccc} W & & \\ \downarrow & \searrow & \\ U & \to & V, \end{array} \tag{33}$$

the vertical morphism from W to U in the above diagram (33) is an element of Hom_C (W, U). We can compose this vertical morphism with the horizontal morphism from U to V in diagram (33). The composed diagonal morphism is from W to V, i.e., an element of Hom_C (W, V). Hence, we have obtained

$$Hom_C(W, U) = U^{\wedge}(W)Hom_C(W, V) = V^{\wedge}(W)$$

for any object W in C. Namely, for $U \to V$ in (31), we have obtained U^ → V^, which is (32), i.e., functor ^ is covariant.

Definition.
This embedding functor ^ is said to be the *Yoneda embedding* from C to C^.

We will describe *the Zen of Yoneda* based on Yoneda Embedding and the Yoneda Lemma. First, let us consider the Yoneda Lemma.

Let C be a category and let F be a contravariant functor from C to the category (*Set*) of sets. That is, for an object U in C, we have a set $F(U)$, i.e., an object in (*Set*). Then, covariant Yoneda functor ^ takes object U to a contravariant functor U^ which is an object of C^. Since both U^ and F are objects in C^, we can consider the set of morphisms from U^ to F in C^, which we write as $Hom_{C^\wedge}$ (U^, F).

Yoneda Lemma In category (*Set*) there is an isomorphism between $Hom_{C^\wedge}$ (U^, F) and $F(U)$ as sets.

Proofs are given in [Michel (1965)], [Hilton-Stammbach (1971)], [Schubert (1972)], [Kashiwara-Schapira (2006)], or [Kato (2006)].

Remark We will begin a comment on the Yoneda embedding functor ^ from C to C^. Since U^ = Hom_C (-, U) is an object of C^, we can apply the Yoneda embedding functor ^ to $F \in Ob$(C^). Then we obtain F^ = $Hom_{C^\wedge}$ (-, F) in (C^)^. As we have U in C and F in C^, we have U^ in C^ and F^ in (C^)^. Then we can write $Hom_{C^\wedge}$ (U^, F) as F^(U^). Namely, we have a lifted version from

U in C and F in C^, where $Hom_{C^\wedge}$ (U^, F) $\approx F(U)$

to

U^ in C^ and F^ in (C^)^, where $Hom_{C^{\wedge\wedge}}$ (U^^, F^) $\approx$ F^(U^).

Therefore, we can restate the Yoneda Lemma as follows.

Restated Yoneda Lemma There is an isomorphism in (*Set*):

$$Hom_{C^{\wedge\wedge}}\left(U^{\wedge\wedge}, F^{\wedge}\right) \approx F^{\wedge}\left(U^{\wedge}\right) \qquad \text{(Y^)}$$

Remark Then we have the following by repeated use of the above lifted version (Y^) of the Yoneda Lemma. As sets, we have isomorphisms

$$\begin{aligned} &\cdots\cdots \\ &\approx \cdots\cdots \\ &\approx \mathrm{Hom}_{C^{\wedge\wedge}}\left(U^{\wedge\wedge}, F^{\wedge}\right) = F^{\wedge\wedge}\left(U^{\wedge\wedge}\right) \\ &\approx \mathrm{Hom}_{C^{\wedge}}\left(U^{\wedge}, F\right) = F^{\wedge}\left(U^{\wedge}\right) \\ &\approx F(U). \end{aligned}$$

Note Here is a psychologically interesting effect-vision of the Yoneda Lemma, which we call *the Zen of Yoneda*. For functor F^ from C^ to (*Set*), i.e., an object of C^^, we have set F^(U^) = $Hom_{C^\wedge}$ (U^, F) $\approx F(U)$ by the Yoneda Lemma. Then at this new level C^^, the Yoneda Lemma implies that F^^(U^^) is isomorphic to F^(U^), which is isomorphic to $F(U)$ at the original level. In other words, the Yoneda Lemma is a level independent assertion, especially when one identifies an object with the embedded object. Continue these processes, e.g., F^^^(U^^^) .... Then, since ^

is an embedding functor, category C can be viewed as an embedded subcategory of C^. That means that an object U in C can be viewed as an object U^ in C^, also as an object U^^ in C^^, etc. If we identify U with U^ by the embedding theorem, we have reached the tautological statement which says '$F(U)$ is isomorphic to $F(U)$.' The following commutative diagram (34) clarifies and summarizes the above lifted vision of the Yoneda Lemma.

$$
\begin{array}{lll}
\vdots & & \\
\wedge\uparrow & & \\
\mathrm{C}^{\wedge\wedge} & & \\
\wedge\uparrow & \searrow_{F^{\wedge\wedge}} & \\
\mathrm{C}^{\wedge} & & \\
\wedge\uparrow & & \searrow_{F^{\wedge}} \\
\mathrm{C} & \underset{F}{\rightarrow} & (\mathrm{Set}).
\end{array} \qquad (34)
$$

Namely, the commutativity of diagram (34) says that the restriction of F^ to C is isomorphic to F, i.e.,

$$F^{\wedge}|C \approx F. \qquad (34|C)$$

We can further ask the meaning of this tautology in terms of temporal topos, consciousness, and fundamental awareness.

The above tautological vision, which we call *the Zen of Yoneda*, i.e., a vision *based on the Yoneda Lemma* and *Yoneda Embedding*, can be considered as one of the fundamental pillars in the category-topos method. Furthermore, we can consider that *the Zen of Yoneda* is one of the fundamental elements in Grothendieck-style-mathematics. See Barry Mazur's article in *Alexandre Grothendieck*, Notices of the AMS, page 404, Volume 63, Number 4, April 2016. To be even more Zen-like, whenever you are given an object of any category whatsoever, we have arrived at the following *realm of Yoneda*, based on the Yoneda Lemma and Yoneda Embedding, which is:

$$\textit{think of an object as a functor} \qquad (34\text{realm})$$

meaning objects of a category C and functors in C^ in the same realm in C^, C^^, etc., by Yoneda Embedding.

Based on (34_{realm}), if one wants to know about object X in C, consider all the morphisms $\{X(Y)\}$ from the objects $\{Y\}$ to X as the characterization of the associated contravariant functor X^, viewed in light of the Zen of Yoneda. Namely, we have

$$X(Y) \approx X^{\wedge}(Y) \approx Hom_C(Y, X). \qquad (35)$$

This view may give us a new vision to grasp mathematics and even how to see *an awakened reality* for an initial object α of category S^ in our temporal topos theory. The reader may notice that for a conscious entity as an object of a temporal topos, one may choose an arbitrarily lifted level category where such an entity is embedded.

Sheaf

Let us recall that a presheaf is a contravariant functor from category T associated with a topological space to the category (*Sets*) of sets. Hence, we need to explain what the objects and morphisms in category T are. Objects of category T are the open subsets U, V, W, … in a topological space. The set of morphisms from U to V, which we write as $Hom_T(U, V)$, is defined as follows. We have $Hom_T(U, V) = \emptyset$, the empty set, if U is not contained in V, i.e., $U \not\subset V$. For the case $U \subset V$, the set $Hom_T(U, V)$ consists of one element, i.e., $Hom_T(U, V) = \{$the inclusion mapping $\iota: U \to V\}$. The inclusion mapping $\iota: U \to V$ is defined by $\iota(x) = x \in V$ for any element x in U. Then, the set $Hom_T(U, V)$ has ι as the only element.

We now have category T induced by a topological space. Then a presheaf is defined as follows.

Definition
A presheaf F is a contravariant functor from T to (*Sets*), which is written as

$$F: T \to (Sets).$$

Example Let R^2 be the category induced by the topological space of the usual *xy*-plane $\mathbf{R}^2$. Let F: $R^2 \to$ (*Sets*) be a contravariant functor defined by $F(U) = \{$all continuous functions on an open subset U of $\mathbf{R}^2\}$. Notice that if U is contained in V, i.e., we have the inclusion mapping $\iota: U \to V$. Then, for

$$\iota: U \to V,$$

in R^2, we have the induced restriction mapping, which is $F(\iota)$ in our earlier notation,

$$r^V{}_U: F(V) \to F(U) \tag{36}$$

defined by $r^V{}_U(f_V) = f_U$, where f_V indicates f is defined over V.

For example, let $f(x) = x^2$ be defined on an open subset $V = (2, 5) = \{$all the real numbers x satisfying $2 < x < 5\}$. Let $U = (3, 4) = \{$all the real numbers x satisfying $3 < x < 4\}$. That is, we have $U \subset V$. Then, the restriction of function $f(x) = x^2$ from V to U, written as $f(x)|_U$, is simply $f(x) = x^2$, where x is in U. We write the restriction of function $f(x) \in F(V)$ from V to U as $f(x)|_U = r^V{}_U(f) \in F(U)$.

Definition

A presheaf F on T is said to be a *sheaf* when the following *gluing*-conditions are satisfied: for a covering $\{V_j\}$ of V, i.e., the union of all the V_j covers V, i.e., equals to V, where j is in an index set J. Let V_{ij} be the intersection of V_i and V_j. Since V_{ij} is contained in both V_i and V_j, we have the restriction mappings

$$r^{Vi}{}_{Uij}: F(V_i) \to F(U_{ij}).$$

Gluing Condition: If $r^{Vi}{}_{Uij}(s_i) = r^{Vj}{}_{Uij}(s_j)$ holds for i and j in J, which means the restrictions from both $s_i \in F(V_i)$ and $s_j \in F(V_j)$ to their overlap U_{ij} coincide, then there is a unique global s in $F(V)$ whose restrictions satisfy $r^V{}_{Uj}(s) = s_j$ for all j in J. Then we say that a presheaf F is a *sheaf* over T.

Namely, a presheaf over a topological space is a sheaf if a given local data provides a unique global element which is obtained by the gluing process. The reason we need a more general concept of a site for temporal topos is the following. In the case of category T, which is induced by a topological space, we recall that $Hom_T(V, U)$ is a set that consists of one element, i.e., the inclusion map only for the case $V \subset U$. For temporal topos, we need more elements in the set of all the morphisms from V to U. A category with a Grothendieck topology is an ultimate space concept over which a sheaf theory can be developed.

Definition

A *Grothendieck topology* on a category S is given by a class $Cov(V)$ for an object V of S, and each element $(V \leftarrow V_i)$, $i \in I$, of $Cov(V)$, is called a *covering family* of V. Such a Grothendieck topology must satisfy the following conditions.

(s. 1) An isomorphism $V \leftarrow U$ is a covering family consisting of one morphism $(V \leftarrow U)$ $Cov(V)$.

(s. 2) For $(V \leftarrow V_i)$ in $Cov(V)$ and for any $V \leftarrow W$ in S, the fiber product $(W \leftarrow V_i \times_V W)$, $i \in I$, is a covering family of W.

(s. 3) Let $(V \leftarrow V_i) \in Cov(V)$ and also let $(V_i \leftarrow V_{ij}) \in Cov(V_i)$, then by composing morphisms, we have $(V \leftarrow V_{ij}) \in Cov(V)$.

Definition

A *site* is a category with a Grothendieck topology. Then, in terms of a site, a presheaf F over a site S with covering families $\{(V \leftarrow V_i), i \in I\}$, $V \in$ S, is a *sheaf* when the following sheaf conditions are satisfied.

Sheaf Axiom: For $F(\rho_i)$: $F(V_i) \to F(V_{ij})$ and $F(\rho_i)$: $F(V_j) \to F(V_{ij})$, where ρ_i: $V_{ij} \to V_i$, ρ_j: $V_{ij} \to V_j$ and ρ: $V_i \to V$, if $F(\rho_i)(s_i) = F(\rho_j)(s_j)$ in $F(V_{ij})$, there exists a unique s in $F(V)$ so that $\rho(s) = s \circ \rho = s_i$ in $F(V_i)$ for all $i \in I$.

Notice that s: $V \to F$ and s_i: $V_i \to F$ for $i \in I$, are based on *the Zen of Yoneda* interpretation of, e.g., for s: $V \to F$ as $s \in F(V) \approx Hom_{S^\wedge}(V, F)$ and s_i: $V_i \to F$ as $s_i \in F(V_i) \approx Hom_{S^\wedge}(V_i, F)$, respectively. Notice also that $F(\rho_i)(s_i) = s_i \circ \rho_i$ is the composition of morphisms

$$\rho_i: V_{ij} \to V_i \text{ and } s_i : V_i \to F.$$

Consequently, we have $s_i \circ \rho_i = F(\rho_i)(s_i)$: $V_{ij} \to F$ which is a morphism in $F(V_{ij})$. See (21) and (22) in A.1.2.

Remark *For mathematically inclined readers*, we can express the above sheaf condition as follows:

$$F(V) \xrightarrow{F(\rho)} \Pi F(V_i) \xrightarrow{F(\rho_i),} , \xrightarrow{F(\rho_j)} \Pi F(V_{ij}),$$

where the meaning of the sequence with morphisms $F(\rho)$, $F(\rho_i)$, and $F(\rho_j)$ are explained in the above.

Final Note If one is *mathematically inclined*, such a reader may recognize with joy, based on *the Realm of Yoneda*, how our morphism-valued interplay between categories and sheaves is ideally formulated and works in a cogwheel-like way in the above presentation.

For a more thorough treatment of sheaves, see, [Artin (1962)], [Makkai-Reyes (1977)], [Tamme (1994)], [Gelfand-Manin (1996)], [Kashiwara-Schapira (2006)], or [Kato (2006)].

We needed to prepare sheaf-category methods to be able to discuss the interplay of cause-effect involving measurement phenomena.

Final Remarks on Cause-Effect and Measurement

We have given a physics reason in Chap. 3 why we use a contravariant functor rather than a covariant functor. The choice of a contravariant functor is also mathematically natural in the following sense. For a temporally linear morphism $f: V \to W$ in category C, a contravariant functor q takes $f: V \to W$ to $q(f): q(W) \to q(V)$.

In *the realm of Yoneda,*

$$\textit{consider an object as a functor.} \tag{34realm}$$

Consider the following diagram in the realm of Yoneda, i.e., based on Yoneda embedding and Yoneda lemma:

$$\begin{array}{ccc} & q & \\ {\scriptstyle s\circ f}\uparrow & \nwarrow {\scriptstyle s} & \\ V & \underset{f}{\to} & W. \end{array} \tag{37realm}$$

Then notice that morphism $s: W \to q$ in $q(W) \approx Hom_{\mathrm{C}}(W, q)$ can be composed with the horizontal morphism $f: V \to W$ in (37 $_{realm}$) as composed $s \circ f \in Hom_{\mathrm{C}}(V, q) \approx q(V)$. Consequently, we obtain the correspondence from $s \in q(W)$ to $s \circ f \in q(V)$, i.e., morphism

$$q(f): q(W) \to q(V),$$

where $q(f)(s) = s \circ f: V \to q$. See diagram ($37_{realm}$). Notice that for a covariant functor such naturalness would not be obtained. There may be another philosophical reason beyond the mathematical naturalness which we are not able to articulate.

The next remark is about a non-canonical observation morphism

$$j: p \to q.$$

G. C. Kato, *Temporal Topos Methods for the Philosophy of Natural Sciences*, https://doi.org/10.1007/978-981-96-2420-1

Let us begin with a commutative diagram in *the realm of Yoneda*:

$$\begin{array}{ccc} p & \xrightarrow{j} & q \\ \uparrow & & \uparrow \\ V & \xrightarrow[f]{} & W. \end{array} \qquad (38\text{realm})$$

Then we have diagram (39_{realm}), where vertical morphisms $p(f)$ and $q(f)$ are canonically induced (i.e., functorially induced) by contravariant functors p and q, and where horizontal morphisms $j(V)$ and $j(W)$ are non-canonical observation morphisms:

$$\begin{array}{ccc} p(V) & \xrightarrow{j(V)} & q(V) \\ \uparrow_{p(f)} & & \uparrow_{q(f)} \\ p(W) & \xrightarrow[j(W)]{} & q(W). \end{array} \qquad (39\text{realm})$$

The commutativity of diagram (39_{realm}) can be proved as follows.

Let us begin at the left corner of the bottom of diagram (39_{realm}). Let $s \in p(W)$, i.e., s: $W \to p$ is an element of $Hom_{Y\text{-}ream}(W, p)$ where we intentionally did not indicate in which category W and p belong in the spirit of the *Zen of Yoneda*; category *Y-realm* can be C^ or C^^, etc.

In diagram (38_{realm}), s: $W \to p$ can be composed with the horizontal morphism j: $p \to q$. Then we have $j \circ s$: $W \to q$ belonging to $q(W)$. That is, for $s \in p(W)$, we take $s \in p(W)$ horizontally to the right arriving at $q(W)$ in (39_{realm}). So far, we have defined $(j(W))(s) = j \circ s \in q(W)$. Next let us chase the diagram (39_{realm}) vertically along $q(f)$. We need to define $q(f)(j \circ s) = (j \circ s) \circ f$: $V \to q$, which is an element of $q(V)$. This completes the counterclockwise diagram chasing of (39_{realm}). Next, we diagram chasing clockwise beginning from the left bottom $p(W)$ in (39_{realm}). As before, for $s \in p(W)$, define $p(f)(s) = s \circ f$: $V \to p$, which belongs to set $p(V)$ arriving at the upper-left corner of (39_{realm}). Finally, to arrive at $q(V)$ from $p(V)$, we define $j(V)(s \circ f) = j \circ (s \circ f)$: $V \to q$, which is an element of $q(V)$. Namely, in diagram (39_{realm}), a counterclockwise element is $(j \circ s) \circ f$, and the clockwise element is $j \circ (s \circ f)$, completing the proof of commutativity of diagram (39_{realm}) by the associativity law.

Having proved the commutativity of diagram (39_{realm}), we are ready to make connections between (I) the principle of cause and effect and (II) measurement phenomenon. Let p be the observed and let q observe p. For a temporally linear morphism f: $V \to W$, we assume either case (*i*) W corresponds to the present time, and V corresponds to a past time or case (*ii*) W corresponds to a future time, and V corresponds to the present time.

For case (*ii*), a measurement phenomenon involves a *cause* for the choice of a future object W among the possible choices as shown in the ur-lightcone in Chap. 3.

A requirement for such a future determining object W is to satisfy the commutative diagram (39_{realm}).

Then, for diagram (39_{realm}),

the commutativity corresponds to the principle of cause and effect.

The reader may find the case worthwhile to consider when $V \to W$ is a micromorphism for elementary ur-presheaves p and q in (38_{realm}) and (39_{realm}).

References

Artin, M.: Grothendieck topologies. Harvard University, Notes on a Seminar (1962)

Barbour, J.: The End of Time, The Next Revolution in our Understanding of the Universe. Oxford Univ. Press. ISBN 0-297-81985-2; ISBN 0-19-511729-8 (1999)

Butterfield, J., Isham, C.J.: On the emergence of time in quantum gravity. In: The Arguments of Time, (ed.) Butterfield, J., Oxford University Press (1999)

Butterfield, J., Isham, C.J.: Spacetime and the philosophical challenge of quantum gravity. In: Physics meets philosophy at the planck scale: contemporary theories in quantum gravity, (ed.) Callender, C., Huggett, N., Cambridge University Press (2001)

Callender, C., Huggett, N., (ed.) Physics meets philosophy at the Planck scale, contemporary theories in quantum gravity, Cambridge University Press (2001). https://doi.org/10.1017/CBO9780511612909

du Sautoy, M.: The music of the primes, Harper Collins Publisher (2003)

Eilenberg, S., MacLane, S.: General theory of natural equivalences. Trans. Amer. Math. Soc. **58**, 231–294 (1945)

Frenkel, E.: Love and math, Basic Books, ISBN 978-0-465-05074-1 (2013)

Gelfand, S.I., Manin, Yu. O.: Methods of Homological Algebra, Springer (1996)

Grauert, H., Remmert, R.: Coherent Analytic Sheaves, Grundlehren der Mathematischen Wissenschaften 265, Springer-Verlag (1984)

Grauert, H.: The methods of the theory of functions of several complex variables. In: Miscellanea Mathematica, Springer-Verlag (1991)

Grothendieck, A.: Récoltes et Semaille (Reaping and Sowing), Parts I, II, III, IV, English translation by Childe, PDF Version 18/07/2022 (2022)

Guts, A.K., Grinkevich, E.B.: Toposes in general theory of relativity (1996). arXiv:gr-qc/9610073

Hadamard J.: The psychology of invention in the mathematical field, Princeton University Press (1945)

Heisenberg, W.: Physics and philosophy (The Revolution in Modern Physics), Prometheus Books (1999)

Herbert J.: Shinto—the fountainhead of Japan, Routledge. Herbert, Jean; 佐佐木 行忠『Shintô: at the fountain-head of Japan』Routledge, 2011–00. Webcat Plus (1967). https://webcatplus.jp/book/12011909

Isham, C., Butterfield, J.: Spacetime and the philosophical challege of quantum gravity. arXiv:gr-qc/9903072bl (1999)

Isham, C., Butterfield, J.: Some possible roles for topos theory, quantum theory and quantum gravity. Found. Phys. **30**, 1707–1735 (2000)

G. C. Kato, *Temporal Topos Methods for the Philosophy of Natural Sciences*, https://doi.org/10.1007/978-981-96-2420-1

Isham,C.: Topos methods in the foundations of physics. Preprint, arXiv:qantph/1004.356v1 (2010)

Kashiwara, M., Schapira, P.: Categories and Sheaves, Grundlehren der mathematischen Wissenschaften332, Springer (2006)

Kato, G.: Elemental principles of t-topos. Europhys. Lett. **68**(4), 467–472 (2004). https://doi.org/10.1209/epl/i2004-10234-2

Kato, G.: U-Singularity and t-Topos theoretic entropy. Int. J. Theor. Phys. **49**(8), 1952–1960 (2010). https://doi.org/10.1007/s10773-010-0380-8

Kato, G., Tanaka, T.: Double-slit interference and temporal topos. Found. Phys. **36**(11), 1681–1700 (2006)

Kato, G.: Elemental t.g. Principles of Relativistic t-Topos (Presheafification of Matter, Space, and Time). Europhysics Letters vol. 71, no. 2 p. 172–178 (2005). https://doi.org/10.1209/epl/i2005-10088-0

Kato, G., Nishimura, K.: Grasping a concept as an image or as a word—a categorical formulation of visual and verbal thinking processes. J. Sci. Res. Reports **2**(2), 682–691 (2013)

Kato, G., Nishimura, K.: An integrated brain function -sheaf theoretic approach to brain as a conscious entity. Annals Cognitive Sci. no 2, 39–43, The Scholarly Pages, 2017, December (2017)

Kato, G., Nishimura, K.: Toward a descent theoretic formulation for organization and emergence—An initial object sheaf hypothesis and its consequences. J. Int. Creat. Stud. No. 2020-004-a (2020). https://doi.org/10.13140/RG.2.2.30759.37282

Kato, G., Nishimura, K.: Time and mnemonic morphism. Creative complex systems, pp. 113–120, Springer (2021). https://doi.org/10.1007/978-981-16-4457-3_8

Kato, G.: The Heart of Cohomology, Springer (2006)

Kato, G.: Sheaf theoretic formulation of entanglement. Quantum Biosystems (2008)

Kato, G.: Microcosm to macrocosm via the notion of a sheaf (Observers in terms of t-topos). Physics of emergence and organization, World Scientific, pp. 229–232 (2008). https://doi.org/10.1142/9789812779953_0008

Kato, G.: Elements of Temporal Topos, Theschoolbook.com (2013)

Kato, G.: Topos theoretic approach to space and time. In: Space, time and the limits of human understanding, (ed.) Wuppuluri, S., Ghirardi, G., Springer, pp. 313–326 (2017). https://doi.org/10.1007/978-3-319-44418-5_25

Leibniz, G.: The Monadology, Kshetra Books (2016). ISBN: 978-1539632603

Leray, L., "L'anneau spectral et l'anneau filtré d'homologie d'un espace localement compact et d'un application continue. J. Math Pures Appl. **29** (1950)

Makkai, M., Reyes, G.: First order categorical logic: model-theoretical methods in the theory of Topoi and related categories, Springer (1977)

Mallios, A., Zafiris, E.: Differential sheaves and connections, a natural approach to physical geometry, series on concrete and applicable mathematics vol. 18, World Scientific Publishing Co. Pte. Ltd. (2016)

Mallios, A., Raptis, I.: Finitary, causal and quantal vacuum Einstein gravity. Int. J. Theor. Phys. **42**, 1479 (2003). gr-qc/0209048

Mitchell, B.: Theory of categories, Academic Press (1965)

Oka, K.: Sur quelques notions arithmétiques. Bulletin de la Société Mathématique de France **78** (1950)

Penrose, R.: The road to reality. Knopf, Alfred A (2005)

Penrose, R.: Shadows of the mind, Oxford University Press (1994)

Quatum Fields and Strings: A course for Mathematicians, (ed.): Deligne, P., Kazhdan, D., Etingof, P., Morgan, J.W., Freed, D.S., Morrison D., Jeffery L.C., Witten, E., vol. 1 and 2, AMS and IAS (1991)

Ranganathananda, S.: The message of the Upanisads, Bharatiya Vidya Bhavan, Bombay, 1993. Universal message of the Bhagavad Gita, vol 1, 2, and 3, Advaita Ashrama, Calcutta (2000)

Raptis, I.: Finitary-algebraic 'resolution' of the inner Schwarzschild singularity. Int. J. Theor. Phys. **45**, 79 (2006). gr-qc/0408045

Raptis, I.: A heuristic philosophical discourse on various applications of abstract differential geometry in quantum gravity research. Int. J. Theor. Phys. **63**, 270 (2024)

Schubert, H.: Categories, Springer (1972)

Smolin, L.: Three roads to quantum gravity, Basic Books (2001)

Tamme, G.: Introduction to Etale Cohomology, Springer-Verlag (1994)

The Leibniz-Clarke Correspondence, Manchester University Press (1956)

Van Oystaeyen, F.: Virtual topology and functor geometry, Chapman & Hall (2008). ISBN 978-1-4200-6056-0

Weyl, H.: Mind and nature: selected writings on philosophy. Princeton University Press. Edited and with an introduction by Peter Pesic, Mathematics and Physics (2009)

Weyl, H.: Philosophy of mathematics and natural science. Princeton University Press, 1 edition. 2 end, Princeton University Press, 1949. 2009 edition, with a new introduction by Frank Wilczek. An expanded English version of Philosophie der Mathematik und Naturwissenschaft, München, Leibniz Verlag (1927)

Witten, E.: Magic, mystery, and matrix, mathematics: frontiers and perspectives, international mathematical union (IMU), (ed.) Arnold, V., Atiyah, M.,Lax, P., Mazur, B. (2000)

Zeami; *Fushikaden Kakyo* 風姿花伝・花鏡 (Essence of *Noh*, originally written and completed in 1424), Tachibana-shuppan (2012)

Index

G. C. Kato, *Temporal Topos Methods for the Philosophy of Natural Sciences*,
https://doi.org/10.1007/978-981-96-2420-1

The manufacturer's authorised representative in the EU is Springer Nature Customer Service Centre GmbH, Europaplatz 3, 69115 Heidelberg, Germany. If you have any concerns regarding our products, please contact ProductSafety@springernature.com

Printed and bound by CPI Group (UK) Ltd, Croydon, CR0 4YY
30/09/2026
02225310-0002